Diary of a Keen Gardener

Also by Mary Keen

The Glory of the English Garden
The Garden Border Book
Decorate Your Garden
Colour Your Garden
Creating a Garden
Paradise and Plenty

Diary of a Keen Gardener

MARY KEEN

JOHN MURRAY

First published in Great Britain in 2025 by John Murray (Publishers)

1

Copyright © Mary Keen 2025

The right of Mary Keen to be identified as the Author of the Work has been asserted by her in accordance with the Copyright, Designs and Patents Act 1988.

All rights reserved. No part of this publication may be reproduced, stored in a retrieval system, or transmitted, in any form or by any means without the prior written permission of the publisher, nor be otherwise circulated in any form of binding or cover other than that in which it is published and without a similar condition being imposed on the subsequent purchaser.

A CIP catalogue record for this title is available from the British Library

Hardback ISBN 9781399829137
ebook ISBN 9781399829151

Typeset in Bembo MT by Hewer Text UK Ltd, Edinburgh
Printed and bound in Great Britain by Clays Ltd, Elcograf S.p.A.

John Murray policy is to use papers that are natural, renewable and recyclable products and made from wood grown in sustainable forests. The logging and manufacturing processes are expected to conform to the environmental regulations of the country of origin.

Carmelite House
50 Victoria Embankment
London EC4Y 0DZ

www.johnmurraypress.co.uk

John Murray Press, part of Hodder & Stoughton Limited
An Hachette UK company

The authorised representative in the EEA is Hachette Ireland, 8 Castlecourt Centre, Dublin 15, D15 XTP3, Ireland (email: info@hbgi.ie)

Charlie and I were married for fifty-nine years. He gave me a life that I never dreamed was possible. Both gregarious and self-sufficient, he had the rare gift of contentment. His presence has guided my life and my writing.
This book is dedicated to him.

Herb Paris

Contents

Introduction	1
November	5
December	27
January	41
February	55
March	69
April	89
May	103
June	119
July	133
August	145
September	157
October	171
A List of the Plants I Currently Grow	185
Acknowledgements	241
Credits	243

A few hours looking at the way light comes through an iris leaf or breathing the scent of wallflowers on a warm morning creates a hole in time: a place we might return to when everything else has fallen apart.

Ben Dark

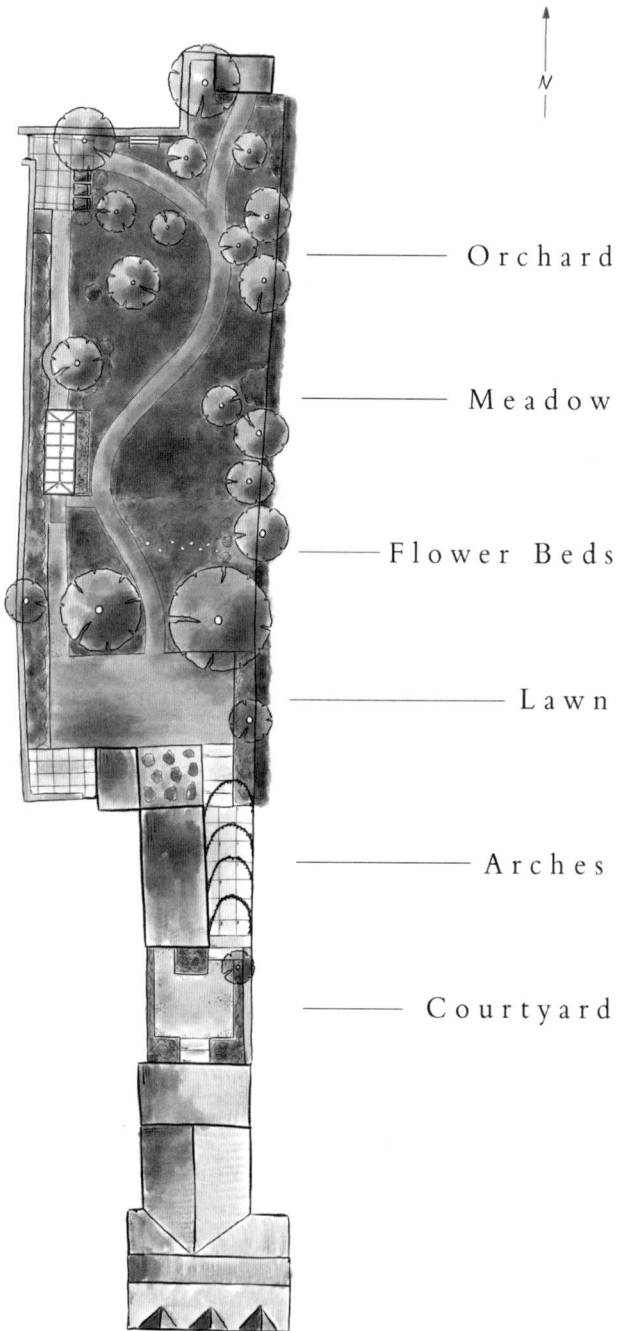

Introduction

'What's your favourite garden?' people ask. And when I answer, 'Mine, of course,' they look surprised, because six years ago we downsized from a much larger garden, which was regularly open to groups from all over the world, to a plot roughly the size of two tennis courts, end on end. Although I have worked on plenty of large places for high-profile people, work at home is what I have always most enjoyed. '#LoveMyJob' is a regular designer boast on Instagram. I do love my job, which has been stimulating, exasperating, rewarding and exhausting by turns, but even the best client and a design scheme that pans out perfectly can collapse if there is no one left to watch over the plants as they grow.

However good they are, designers and photographers can struggle to conjure one of those heart-stopping places where the lives of plants and people seem interwoven. Because it's only the constant observation, the hands in the earth work, which creates a place to be rather than just somewhere for show. My eldest daughter, Laura, says that the Greek word for gardener means 'garden watcher'. It's the daily watchfulness, the constant commitment, that creates a garden. The late John Sales of the National Trust used to say that a garden is forty per cent design and sixty per cent maintenance, but designers still get more coverage than maintainers. So, this diary sets down something of what it is like to be a lifelong gardener.

It begins at the lowest ebb of the gardener's year, when leaves are falling and flowers are retreating. Time slows down in November, but the way time changes is all part of gardening. It can slow to a standstill as you stop to stare at a flower, or speed to a flash of the past, all while the seasons just go on turning in their same comfortable way. I wanted that timelessness to be a part of this book as well as a celebration of the ordinary, the ephemeral and the useful. So, the diary is a patchwork of thoughts and tasks and life, from past and present. It's not about celebrity or display but about the community of those who grow. There is a little about where I live and a lot about what I do, about what I have learned and am still learning about plants. Alice, my third daughter, another classicist, tells me that *oikos* is the Greek for 'home', and *logy* means 'written expression', which adds up to 'ecology'. Ecology begins at home.

Home is a village at the unfashionable end of the Cotswolds, with a faintly fifties feel. On our first visit, a man who might have been an extra from an Agatha Christie film was bicycling down the high street in a straw hat. People greet each other as they pass and there is always much stopping and chatting outside the shop. The area within a ten-mile radius has attracted several gardening friends. Dan, Hannah and Catherine are designers; Derry runs a brilliant nursery, as do Rob and Gussy; Jo and Alison are new wave eco missionaries who teach others regenerative gardening; Jonny is a peripatetic gardener and writer; and my three daughters, Laura, Ellie and Alice, also local, are all gardeners who have, in hard times, been employed as gardeners. Even my actor son William has worked as a gardener in periods of 'resting'. Instagrammers from the wider world are another source of inspiration and advice. Tom, Sarah, Arthur, Fergus, Julian, Isabel and Tania are friends and horticultural stars whose names will be familiar to committed plant nuts. Pip Morrison was my working partner over a period of twenty years and remains a good friend. We exchange information with each other and with other hands-on growers who are similarly obsessed. Most days I learn something new about plants.

NOVEMBER
Mistletoe

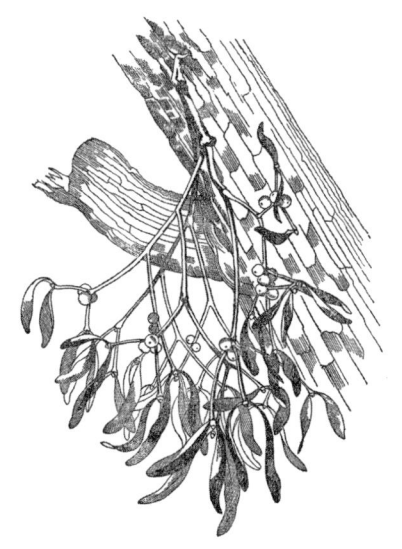

Today, Juliet, a Sussex friend, posts a picture on her Instagram of a road which is now a river, but at home the sun is shining and the sky is blue. November, and so far we have had no frosts. Most nights I have not bothered to shut the greenhouse, but the garden dwindles, shrinking at the prospect of winter. Spots of colour still flash from salvias: 'Cerro Potosi' is shocking pink, and there are a few brightest-blue hooded flowers on the giant form of *Salvia patens*. Dahlias 'Winston Churchill' and 'Karma Fuchsiana' carry on being outrageously bright. In the courtyard, drooping sprays of salvia 'Phyllis' Fancy' attract hummingbird moths and the climbing fuchsia 'Lady Boothby' flowers and flowers, so suddenly I want to know who Lady Boothby was. Was she a gardener? She is hard to find. The spook in my iPad thinks I am only interested in flowers, so it serves up pictures, shopping opportunities and advice about growing fuchsia 'Lady Boothby', but nothing about the Lady herself. I try Lord Boothby, whose life was as colourful as the gaudiest fuchsia. Maybe his second wife, Wanda Sanna, who was Italian and half his age, was a gardener? In a painting of her in the National Portrait Gallery she is looking glamorous and a bit anxious, perhaps because her husband is said to have been a friend of gangsters and a famous philanderer. Now I wonder if I like my fuchsia less.

The provenance of flowers is a part of why we grow them. Gardeners love flowers which remind them of friends or family or holidays. I tolerate the silver-leafed weeping pear which we inherited here, although it was a cliché of mid-twentieth-century gardens, along with purple leaves and ground cover plants. Fashions change; shrubs with coloured or variegated leaves were popular in the fifties and sixties, and were then rejected. Recently, there has been a shift to shrubs again.

My mother-in-law, Catherine, who encouraged me to garden, was an inspired hands-on artist gardener who planted with her Slade-trained eye. Ribena-coloured 'Rosemary Rose' and acid-green *Alchemilla mollis* was a daring combination at a time when most others were choosing pastel shades. But there was a silver-leafed pear in Catherine's garden. In the sixties, silver foliage was made fashionable by Mrs Underwood, owner of a nursery dedicated to growing only plants with silver leaves. I can just remember her at Chelsea shows, where she was gracious and condescending to a young gardener like me. But in the biography of Mrs Underwood, written by Catherine Horwood, she sounds more frightening than in my memory. Tall with a booming voice and usually a cigarette dangling from her lips, Mrs Underwood has been described as 'a woman of high standards and great courage', a euphemism for being rather terrifying in an English country lady sort of way.

There was plenty of silver in the Oxfordshire garden of my in-laws, but very few pastel shades, apart from a six-foot pink floribunda 'Queen Elizabeth' rose (a plant that's fallen out of fashion now). Their garden was never what my friend Margaret would refer to as 'quiet good taste'. Quiet good taste dominated provincial life, even after the sixties changed the way young people like us lived in the cities. We were away from the bright lights, dressing sensibly, raising children in the Midlands. Sometimes we escaped to see what our contemporaries were doing. 'Have you solved the awful dichotomy?' one woman asked. What she meant was, how did I manage with nothing but home and children to occupy me? At the time I did not have an answer. It was lonely being away from friends with only two tiny girls for company, but Catherine saw that I needed something to occupy my spare time, which would fit with looking after small children, so she began to teach me about flowers. Pruning roses in the garden of a rented house while eight months pregnant was not what got me hooked. In those days you had to bend to cut to an outward facing bud in the icy days of March. Pruning is much more flexible now, and can even be done with shears or a hedge trimmer pretty well whenever you feel like it. Sometimes I listen to archive editions of *Gardeners' Question Time* and am surprised by how much methods and taste have changed in the last fifty years.

When we moved to our own house and I could choose what to grow, I began to see the point of gardening. 'Very Tennysonian, darling,' Catherine says of the house we have bought in Lincolnshire, but she tells me about old roses being more beautiful than hybrid teas. I read Scott's catalogue in the bath, which is almost a book: a slim volume with line drawings by Robin Tanner, and masses of facts about plants. I plant 'Nuits de Young' and 'Tuscany' and *Rosa gallica* 'Versicolor' under the arched windows, and we hardly mow the lawn because it has daisies all over it and the little girls run through the flowers wearing dresses with roses painted round the hems. Then we lie on the grass and make daisy chains which somebody always breaks, but you can mend them if you make a longer slit to put the flower through. The daisy stalks have a sour taste, but the petals are sweeter. I chew the grass stems, which are sweetest of all at the base, but I don't let the children do that.

Then, years later, in another house where we lived in our twenties, when my baby son cries across the landing, I leave our bed to feed him. After he falls asleep, I begin to go back to bed but the sky outside the landing window brightens as I stand there looking at the river and listening to the early dawn birdsong. It is tempting to go outside for a bit, because soon the baby will wake, and the garden may never look so beckoning again. In wellingtons and a nightie with a coat on top, I pull goose grass out of the peony leaves and think of Miss Archer Houblon who made this garden. She planted wild yellow tulips in the gravel and shuttlecock ferns on the banks of the chalk stream.

The scented wild yellow *Tulipa sylvestris* is a British native, although I have seen it growing under olive trees in France. I have tried to grow it in all the gardens I have made in different homes since then, because it reminds me of our idyllic rented farmhouse by the River Lambourn, surrounded by overgrown relics of Miss Archer Houblon's garden. 'Such a mistake to get off the housing ladder,' the grown-ups said. But we did, and it wasn't. Or it may have been, but the compensation was being able to swim in the icy pool below the waterfall and on summer evenings Charlie could fish for trout. In the July dusk, while Buzz Aldrin and Neil Armstrong were landing on the moon, my best friend Grizelda and I waded upstream, past royal ferns and spent primulas. 'Better than the moonshot, any day,' Gelda wrote in her thank-you letter.

So, *T. sylvestris* is a souvenir from that early idyll. Here, in what may be my last garden, I have put some in the little orchard, but they came up with twisted petals which is most likely tulip fire, the virus that can affect bulbs in badly drained places. I am puzzled, because this garden is on Cotswold brash, and I can grow the red *pavonina* anemones, which hate wet ground. (Most of us are still struggling with their name change to *hortensis*.) Perhaps the bulbs I bought were infected when they arrived. The garden designer Bannermans, who live in the wetlands near Yeovil, next to a village called Mudford, have sheets of healthy *sylvestris* with scented delicious flowers. I think I should order more.

Kew-trained Beth and her artist daughter Tess are staying, so we walk up the road to see Sheila, who makes silver knuckle-duster rings. She is Dan Pearson's mother and has a beautiful tiny garden designed by her son. But as we pass Beth's parked car we notice it has a flat tyre. We are three generations of women stopped and staring at this thing, whose needs we none of us understand, on a sunny Sunday morning when every local garage is closed. A man carrying a newspaper pauses, concerned, to ask if he can help. In this village where I live, people do that. He offers to check the tyre pressure at his house round the corner; Beth can drive it up there, it will be fine. But it is not fine. As Beth is turning the car in the entrance to the pub car park, the rubber comes away from the hub. She cannot drive anywhere. Tess runs round the corner to fetch back the Good Samaritan.

Unnamed but now canonised, he kneels on the damp November ground to remove the tyre with the jack, and we take it back to his house to check it. Spares are a thing of the past, he tells us; but it may just be a slow puncture. We watch him meticulously checking for bubbles of air after he has blown it up. A blonde girl in a car stops to shout hello to him. We say, 'This man is an absolute saint.' When we finally get the tyre back to Beth's car, the GS, who we now know as David, bolts it back on and tells Beth he is sure it is indeed a slow puncture. She should watch the tyre pressure gauge on the dashboard and, if it falls, stop at the next service station to blow it up again. We thank David, who sets off home for the *third* time to have his Sunday lunch, probably cold now, with his family. I love this village and the kindness of strangers.

Last time Alice, the third daughter, came, she turned out the huge compost heap at the top of the garden. The good compost from the bottom she piled under the apple tree, and the unrotted stuff went back. She likes a tough task. At the allotment over the main road (the garden isn't big enough to grow veg), I have three smaller heaps divided by pallets which never produce compost like the stuff which comes from the untended heap in the garden. Large prunings, cardboard boxes, mowings, masses of weeds all get thrown in and, after a year, a crumbly mixture the colour of chocolate cake is ready to spread over cut-down dahlias and around hellebores and roses. So, my unscientific conclusion is that a large heap is the answer to making good compost, because I have tried to be more organised, with minimal success.

I have done the Dowding 'no dig' course – his heaps are huge. I have listened to Bob Flowerdew on *Gardeners' Question Time* and, nearer home, I have had lessons from Peter in the village, our best singer and compost maker. Peter has several bins in the churchyard. On one side Matthew, Mark, Luke and John; on the other Deborah, Ruth, Rahab and Esther. All of them produce enviable friable results. But I know Peter spends hours snipping sticks for layers of brown between the green, while listening to Bach on his headphones. He covers the heaps tenderly so that they are never too wet, and he waters them when they are too dry. He takes the temperature of his compost and when it is cooked, he loads a huge barrow and wheels it down the empty street to needy village gardens at £10 a time for charity. His compost is unrivalled, but life is too short for all that snipping and layering, so careful compost is never going to be one of my skills.

Younger friends swear by the Japanese bokashi method, which ferments food waste in a cupboard in the kitchen. This involves many rituals of mashing, bran spreading, liquid draining. Regularly. My local designer friend Catherine FitzGerald says, 'It's like feeding another person, worse than feeding the children,' and it sounds far too challenging for me. The hot bin at the top of the garden is fine for kitchen waste, which I am reluctant to put on an open heap as it can attract rats. I do tear up cardboard and paper for the hot bin when I remember, and a layer of grass cuttings speeds up the decomposition if the temperature falls. One hot bin produces a couple of barrows of compost about three times a year. But Alice's heap is ten times bigger than that and I must move it soon, because underneath it there are aconites and snowdrops waiting their turn. No frost still, so I won't cut down the dahlias this week. But next week, halfway through November, it will be time. I leave most dahlias in the ground, covered with a bucket of compost and some evergreen branches, but one or two are dug and stored in the greenhouse, in case of failure.

Last Christmas, I gave my granddaughter Emily an online course taught by Shane Connolly, the royal florist. I admire Shane for his campaign to use home-grown flowers and no oasis, even for massive occasions like the King's coronation. In August, Emily came to stay, and because she was going to do the flowers for a friend's wedding at the end of the month, we practised Shane-type arrangements on the kitchen table using flowers and branches from the garden and allotment. That was such a success that she was asked to do more flowers and now she has given up her career in filmset design to become a florist. Today she texts me to say she has a wedding next May for her boyfriend's sister, what can she grow in his mother's greenhouse to be ready for that? Blue flowers and *Ammi*? But it's November and I am doubtful, so I write:

> Darling Emily, I've been thinking a bit about your May wedding flowers for Emerald. I've talked to my gardener friend Peter Dennis and neither he, nor I, are convinced love-in-a-mist will be out in time, but it's worth a try in the greenhouse and so too are cornflowers. But because they are hardy annuals, they really need to be in an airy greenhouse with the doors open, and I think if they do look like flowering, you can accelerate them or retard them nearer the time. But I don't think you could rely on growing them for the day.
>
> May is a heavenly month, with masses of blossom – apple, hawthorn, crab apple, lilac and cow parsley, which are all just

as good as Ammi. What about bluebells? Again, chancy, they might be over if it's very hot. Then these Camassia esculenta, which were out here on the 20th May last year, could be planted in a row in the kitchen garden. They cost £4 for twenty-four from Peter Nyssen. I think you should ask Lucy V about growing for a date. Her mother may use a fridge to hold things back. Or a polytunnel with warm air to speed things up. The weather is so unreliable it's never easy to predict what will be out when. But there will be something, so you may have to adapt. Blue aquilegia granny's bonnets are usually out. I recommend the camassias anyway and give the love-in-a-mist and cornflower seeds a go in a cool greenhouse. Of course, if mine are out in the allotment you can use them. They seed themselves much too prolifically. I've even got some flowering now.

I call in for tea in Buckinghamshire with Jonathan Cooke and Sue Dickinson, legendary gardeners who are now retired from one of my happiest commissions. I ended up writing a book about the way it was gardened called *Paradise and Plenty*, because it seemed important to record a way of cultivation that is almost over. Huge walled gardens, where fruit, vegetables and flowers are grown in double-dug rows to supply a grand house, are increasingly rare. Modern crops are raised in much less labour-intensive ways.

Last week Jonathan and Sue arranged home-grown flowers from the village for a memorial service, using two hundred nerines grown by a neighbour in nearby open ground. Their house is brimming with vases of hardy chrysanthemums, so we examine them all in detail. I like the ones with pronounced yellow middles less than the soft doubles. Sue likes a yellow form of 'E. H. Wilson'. These two, who know everything horticultural that there is to know, are excited by *Which? Gardening*, for its methodical and useful advice, so I leave with a copy and of course they are right. If its frills-free tried and tested research is good enough for two of the very best gardeners in England, I should get it. More than that, it has columns by some of my favourite horticulturalists: Fergus Garrett, Bob Brown and Ken Thompson. How come I have missed this? I should definitely subscribe.

Next morning, I look in on the Paradise and Plenty garden where, since my original design, I have made regular advisory visits for over thirty years. Daniel Jones, the new young gardener there, wants to replace the half-hardy bedding under the old-fashioned roses with perennials. When I ask with what, he says *Geranium* 'Cloud Nine', which I don't know. It's a pale blue double meadow cranesbill which he thinks will flower for longer than 'Mrs Kendal Clarke'. The *Alchemilla mollis* is being replaced with *A. conjuncta*, which has pretty margined leaves and is smaller and less invasive than *mollis*. We look at the hardy chrysanths, in the ground this time rather than in vases. They are such brilliant flowers for this latest season of the year, but hard to include in borders, because they can look rather stiff, so mine are grown at the allotment. My favourites are 'E. H. Wilson', honey-scented, which lasts three weeks indoors, and I love 'Dixter Orange' and 'Emperor of China', with their twisted pink quills for petals. Daniel likes pale yellows and the spider chrysanths, which I grow in the greenhouse, but they seem fine out of doors in the

Buckinghamshire garden. Even my hardy forms sometimes fail to return in spring, so I lift a few favourites to be sure of enjoying them next autumn and store them on the lowest bench. My greenhouse space is minimal, but in this, the grandest of gardens, there are five ninety-foot-long pit glasshouses, so there is room to pot ten of each variety and bring them indoors to a cold house as soon as frosts start. They will go on flowering and when they finish they are cut down. In spring, when the shoots start to grow, cuttings are taken. But at home I find I can make divisions of two or three rooted stems from my potted plants. They need more light from about February, but by then the bulbs on the upper shelf will be over and there is more growing space. Slugs love the new shoots of chrysanths, so I try not to put them out until they are about six inches high. Gardening is a roller coaster. Even in the darkest months, you can look forward to flowers next year, just as long as you keep checking and protecting your plants from weather and pests.

Cut flowers are a major feature in the Paradise and Plenty garden. We pass clumps of healthy delphiniums, which Daniel says have been marvellous this year. He has heard that a cold winter is what delphiniums need. I never knew that. Not sure if it is true, but this running exchange of triumphs and failures with working gardeners is what matters.

To cut or not to cut is the question at this time of year. Dan, in the valley not far from me, leaves everything until spring. I am not sold on brown, which is rarely appealing in soggy English winters. Rimed with frost in northern European perennial borders, things might be different. Anyway, I have snowdrops and primroses and *Cyclamen coum* under the summer flowers, as well as scillas and hellebores, and I want to make sure they have space to emerge. Dahlias, like Cinderella at midnight, turn to rags and tatters as the clock strikes frost, so they will be cut to the ground, then marked and mulched with Alice's excavated compost. I will dig up one of each of the ones I like best, and hope that the rest survive. If they fail, cuttings from the stored plants can be taken in spring.

The rest of the cutting down is for me a matter of choice. *Iris sibirica* can stay, because the dark seed pods are handsome. Campanula of the mushy stems can't. Asters, now horribly renamed *Symphyotrichum*, can stay, so can the skeleton stems of *Althaea cannabina*. Phlox can't, bronze fennel can't. But on my way home I look in at Worton kitchen garden, to find beautiful flowering stems of common fennel (*Foeniculum vulgare*) in a vase on the cafe table. Had I cut mine down in August to rise again, it would still be looking good now.

About the giant burnet, I am undecided. The black bobble seed heads are fine, but the wet brown leaves clinging to the stems look horrible. It all depends on how long they cling. Russian sage, lovely *Laserpitium siler* with giant umbel seed heads, dangling dieramas and grasses, all get winter passes. There are hips on the *Rosa moyesii* and magnificent lacquered ones on *R.* 'Scarlet Fire', and once the leaves drop, red stems of the dogwood will show. A few *Cyclamen coum* have started to flower, but apart from green, it's goodbye to colour now until next year. So, green matters for the next few months when the background players come into their own. A big bush of *Bupleurum fruticosum*, which I have to prune or else it sits on everything; a fastigiate box grown from a cutting taken from one of the four bushes of *Buxus*

'Graham Blandy' in our last garden; some *Sarcococca*, some cistus, euphorbias, daphnes, ferns, especially 'Bevis', outside the kitchen door. These will be my winter companions while I wait for snowdrops and hellebores to emerge.

Going out to the garden again, I realise I have ignored the two best of all winter comforters. The winter-flowering cherry (*Prunus* × *subhirtella* 'Autumnalis'), which marks the entrance to the orchard meadow, now has pretty russet-tinted leaves as well as the first buds of blossom, which will appear along its branches until Christmas and beyond. This tree was one of our first gardening presents from Catherine, my artist mother-in-law, in our earliest garden. If it had not been here when we came, I would certainly have bought it. My other Desert Island plant is the starry yellow jasmine (*Jasminum nudiflorum*). I did buy that. It has not started to flower yet, but I have just picked some stems, and they will come out on the kitchen table.

I have been invited to join an eco book club the other side of Bath. My friend Derry Watkins, of Special Plants Nursery, offers to drive us, and we spend the evening at Jo McKerr's beautiful house. The book, written by a Japanese farmer called Masanobu Fukuoka forty years ago, is called *One Straw Revolution*, and is a celebration of natural farming. We hardly talk about it. Instead, we discuss compost and whether the garden at Dixter is as biodiverse as the land surrounding it. Apparently it isn't, which is a pity as I just claimed it was in an article in *Country Life*, although no one has yet contradicted me. Jo is a fierce defender of trees, hedges and untamed gardens. We are at a moment when the wilding versus gardening debate is hot and strong. Monty Don has claimed that rewilding gardens is puritanical nonsense. Alan Titchmarsh has mixed feelings about the trend. But Fergus Garrett at Dixter thinks there is no contest. We can have both. I hope we can, but when I show this garden to people in my village, I can see that they are very doubtful about long grass instead of a nice lawn.

In the afternoon, Laura comes to wheel barrows of Alice's compost while I cut down dahlias and phlox, because it's sunny and this is never an agreeable job in the rain. For gardening after the end of October you need to pick your moment, and this is a good one with a daughter on hand to help. Usually mornings are better than afternoons, but we are lucky and could go on working until teatime. Standing waiting for the next barrowload to arrive, I look up at the apple tree to find a sprig of mistletoe has lodged itself in the bark. How did I miss that? It's quite a big sprig and I've always wanted mistletoe, even tried to rub berries into apple trees in at least two gardens before this one. Suddenly here it is: how did it arrive in this enclosed town garden when I have never seen any growing nearby? Birds are said to bring it, but however it came, I feel triumphant. Not for kissing-Christmas reasons but because this is an ancient Celtic magic plant. When I look it up, the Kew website says, 'Its evergreen leaves and ability to bloom in winter mean that the plant has long been seen as a symbol of eternal life and vitality.' Which is handy as over tea Rupert and Laura help to unpack the blood pressure monitor which has been on the kitchen table for a week, waiting for technical help to arrive.

I hate having my blood pressure taken, and the last few times it has been high. 'Could be white coat syndrome,' my doctor says sympathetically. 'Try taking it at home yourself until you get used to it and see if it improves.' Laura volunteers guinea pig duty and pronounces it extremely painful because nobody read the two fingers under the cuff rule, so it was much too tight. Rupert goes next and his registers an off-the-scale error. My turn and it is definitely uncomfortable. The reading is not reassuring. Rupert does his again and comes out top. 'Take the pills,' he says as they leave. I look up 'side-effects of blood pressure pills' and think I'd rather rely on lack of sugar and luck of mistletoe.

This morning, as I start on my second cup of coffee, there is a knock on the kitchen door. It's Dan cradling a small box, and in it are two perfect Comice pears sitting side by side. Saffron gold with a rosy blush, they are so large I have to measure them. I draw round them carefully and the diameter at the widest part is four inches. 'Eat them quickly,' Dan says, but his mother, Sheila of the silver knuckledusters, is coming to lunch with Trudie, my neighbour, tomorrow, so I put them in the fridge, hoping they will keep for twenty-four hours. 'Doyenné du Comice' likes the sunniest wall, which here I do not have. I grew it at our last place, but never as well as these beauties. 'Concorde', the pear for small gardens, is the nearest I can get to 'Comice'. It is a cross between 'Conference' and 'Comice', and self-fertile, which is perfect if you only have room for one pear. Mine is a tree, at the back of the little orchard, so far not triumphantly productive, but I can wait a bit longer.

Once a month, Jane Barnwell, my gardener friend, comes. I love these days, especially yesterday when it was sunny all the time we worked. When I come in to put the soup on for our lunch, Jackie, who helps indoors fortnightly, says, 'You girls must be crackers, working outside all morning.' 'But we love it,' I say, and, 'we'll be out all afternoon too.'

Jane is a tidier and more methodical gardener than I am, so we bicker a bit about what shall be cut and what shall stay. We plant some bulbs, layers of hyacinths and tulips in the big pots outside the greenhouse. This is annoying as the 'Patons Unique' pelargoniums are still flowering. The tulips can wait, I have often planted them in December, but hyacinths need planting, so the Patons get cut down and are potted and put under the greenhouse shelf. Old books say you can store pelargoniums dry, upside down in a shed, but I tried it once and failed. Ideally the greenhouse would all be cuttings in pots nine centimetres square, which take up less room, but I like some big plants to put on the windowsill at the front of the house. Throughout the day we both keep losing our new yellow Niwaki secateurs, so time is wasted by looking among fallen yellow leaves.

This is the time of year when labels matter. Everything potted, all the cuttings must be labelled. The silver pens are cussed. Sometimes they work, sometimes they fail or blot. I used to think, I can't be bothered, I'll definitely remember what that is. Come spring, memory is untrustworthy, and it's a case of wait and see. We find some pots in the auricula shelter which say 'Not sure' in Jane's writing. They might be scented violets. I do grow a few, but only time will tell. Dahlias need labelling and markers for snowdrops appear. I use a silver pen on black labels. I have tried wooden ones, but the writing never lasts, so I keep the wooden ones for short-term seedlings.

Over lunch I ask how Jane's husband Richard is, and she tells me he has gone to Nigeria for six weeks to sort out the boundaries in Gashaka-Gumti, the national park he was involved in setting up with the World Wildlife Fund during the eighties and nineties. Seventy now, it turns out Richard is the only person who still has the original map of the place. I admire these two who worked in Africa for twenty years before deciding they wanted to bring up their children in England. Jane was a domestic science teacher, but now she gardens occasionally for me and a couple of others, and they have a house in France, where since Richard's eyes started failing, Jane drives their truck all the way to the Jura. She is an impressive woman.

I still go to Robert-the-dentist in Cirencester, which means driving back to the area where we used to live, past birches with glittering tresses and beeches with burnished halos. It's a beautiful day again, but the little road through Shipton Moyne is flooded, and I fear the car will pack up. When I get there, Robert shows me a picture of an early miniature daffodil already out in a stone trough in his garden. A daffodil, at the end of November. I say, 'No daffs in my garden yet, but I do have primroses.' We exchange views on the weather. Conversational opportunities at the dentist tend to be one-sided. As my teeth are being polished and I am gagged, he asks me how my actor son is doing and tells me that he used to be Laurence Olivier's dentist when he worked in London, and that he was once invited to Canary Wharf, where the famous actor lived. 'I danced with a man, who danced with a girl, who danced with the Prince of Wales,' isn't in it. When I get home, I look for pictures of Olivier, to admire Robert's work. But I can only find a couple of smiling ones, with a small hint of teeth. Now everybody in a photograph opens wide. I don't think that was such a thing in the sixties and seventies.

Nobody can resist Derry who organises the talks for the University of Bath Gardening Club, so we get the best horti speakers possible. Tonight, it's Peter Korn from Sweden, who grows plants in sand, washing any soil off their roots, which he then spreads on the ground and covers with more sand. He shows us a slide of pulsatillas with huge root systems dangling. When he plants, bare roots are exposed to sun and wind while they wait to go in the ground. But I get nervous if plants are left with roots in the air and always rush to cover them, or to dunk them in water. The Peter Korn edict is no water, no compost, no staking. Ever. He is such a star: the hall is packed. There are grey-haired gardeners, cool young landscape architects and working gardeners hanging on every word and slide. Occasionally, Catherine and I exchange bewildered glances, and as we leave, Catherine says, 'We've got to unlearn *everything* and start again.' I grab the last copy of his book, but I haven't started it yet and it's too heavy to take to London where I am heading today. My first thought is, this develops the work of Beth Chatto – right plant for the right place – and of Keith Wiley, who has sculpted his garden at the edge of Dartmoor into billows and valleys in order to please his plants (Peter Korn's book is called *Giving Plants What They Want*). Second thought is that this very low maintenance way of gardening might mean that the problem of never being able to find enough good gardeners for clients would be solved. My third thought, which Catherine, also a designer, shares as she drives me home, is that I don't want to give up growing greedy plants – peonies, roses, peppery scented phlox, delphiniums – because I love them.

When I open the book, it turns out there is a *Delphinium geyeri*, a delicate larkspur, but the index lists no roses, no peonies and only cushiony alpine phloxes. The other problem is that we are wedded to mixed borders, to height. Fashionable grasses don't always do it for us, or for our clients. I know this is the way forward for gardening, for the sake of the planet, although like St Augustine who prayed to be chaste, adding, 'But not yet,' it's not quite yet for me. But I must read the book carefully and try harder, to water less, obviously. But no compost?

Supper after the ballet in London with Antonia, and lining her paved path are loosely clipped balls of winter jasmine, an original way of growing the yellow star-flowered winter favourite, which I have never seen before. In China it is known as 'the flower that welcomes spring'. I grow it here, not brilliantly, because if you forget to prune it after flowering for a year, bald patches appear. To prune winter jasmine properly takes time. Each shoot that flowers must be cut back. The trimmed hedgehog approach leaves the jasmine looking too bristly for my taste. It might work around cottage porches, but I enjoy this shrub for its graceful green wands, so I want to be careful. In a way it looks better cascading down a wall rather than climbing up it, because as a climber it needs tying in, but as a faller, the gardener is let off the fiddle of fastening. Vita Sackville-West suggested growing jasmine both up and down a stick wigwam, which sounds pretty, but tricky, and I have also seen it growing through ceanothus. However you grow it, it's a must-have plant for every garden.

DECEMBER
Zonal geranium

So cold, so cold today, and grey. Red-check tablecloth, fingerless gloves and too many pieces of hot buttered toast in order to keep warm. *Salvia* 'Phyllis Fancy' has finally packed up, but *Pelargonium* 'Clorinda' in the big pots outside the kitchen hasn't turned a leaf yet. I wonder if I should cover the mimosa, which was halved last winter, reborn into a tidy bouquet of finely cut leaves in the corner of the courtyard wall.

I didn't bother. Yesterday was one of those rare sparkling days when every plant looks like a Christmas decoration. The umbellifers are best of all, but the silver honesty seed pods, which I admired so much in Cambridge last winter, are not self-supporting here.

My honesty is the Corfu perennial blue form. I brought its seed back from Corfu and, after I failed to raise it, Derry of course succeeded. Perhaps the ordinary biennial *Lunaria annua*, or 'Chedglow' with dusky foliage, would make stronger plants in winter, but their flowers are heavy purple. Too dark for me in the clear light of May, when the blue Greek form can appear almost translucent. All honesties seed freely, but their taproots are tricky to transplant. Arthur Parkinson, who is a genius gardener, says the stems need harvesting in summer, so that they are stronger if you want to use them for indoor decoration at Christmas, but I want them to be strong in the garden. There is no shortage of honesty because it self-seeds. John Sales called this 'the happy accident' that gives a garden frayed edges. Welsh poppies, the posh cow parsley *Cenolophium denudatum*, vivid *Silene coronaria*, primroses, cowslips, hollyhocks, forget-me-nots put themselves everywhere here. When a plant self-seeds, it is happy.

I'm meeting a friend off a train at Chippenham and the train is late. Before I left home I put a fish pie in the oven, and now it will be getting burnt. The front door is open so I ring the shop next door, to ask if someone will go to the house and rescue the fish pie. No answer, so I try Louise opposite, who organises theatrical events from a retired ice-cream van which still has Mr Whippy and the Conettes painted on the side. She is often seen on stilts at village events. 'Of course I will,' she says, and adds that she loves the idea that I can ask and that the front door is not locked.

It is the Christmas wreath on the door moment. If you live on the High Street as I do, around now a slip of paper appears on the doormat, inviting you to enter a competition for the best wreath, to be judged by the WI on Christmas Eve. My wreath is never a winner. The chosen ones are bright bought numbers, or novelty wreaths with teddies, or sweets, or dangling discs. There are silvers and scarlets, but mine is always just home-grown green. Each year I tie little swags of mixed evergreens to a twiggy circle. I'm eyeing Trudie next door's willow and thinking of asking her for some branches to refurbish the circle. There is hazel at the allotment, but it isn't as pliable. The inner ring needs to be strong.

There is plenty of flowering ivy on the walls and I can cut some of my neighbour's overhanging thuya, because it is nice and frondy. I have box, *Sarcococca* and euphorbia, maybe some artichoke leaves if they are still whole after the frost, and a teazle or two, but I am going away to stay with Laura and Rupert in Greece ten days before Christmas, so this year durability is key. One year Pippa, my Somerset friend, gave me some of her mistletoe; another year I wired lemons on to the greenery. If I can find holly berries, it's a good year. Crab apples would be lovely but the fruit on the *Malus hupehensis* here doesn't last. 'Red Sentinel' hangs on longer, but that I do not have. I can add some periwinkle to trail, and a few clusters of scarlet hips, but it won't be showy. Instagram is full of wreaths with peppers, fruit, ribbons and everlasting flowers. Too bad.

Last time the car was MOT'd in the village, and as I was paying my annoyingly large bill, I noticed an enormous money plant (*Crassula sarmentosa*) on a high windowsill in the office. 'Can I take a cutting of that?' I asked, and Nick the mechanic said, 'I've got lots already potted. I keep them for customers. You can have two.' As a boy, my son William used to have a money plant, which he believed would bring him luck. I can't remember what happened to his plant, but I do remember it was a craggy characterful thing by the time he left school. He has just moved into a flat in London at a time when he could do with a bit of a financial boost, so I thought he might enjoy the irony of having another money plant. *Crassulas* are succulents, they need gritty soil and very little water to survive. When I got home, I potted both cuttings into free-draining soil and put one in plastic and the one for William in clay. Of course, I like clay pots better than plastic ones. I have a huge collection of varying sizes, stored upside down on industrial steel shelves in the little paved area next to the potting shed. Terracotta pots being porous drain faster than plastic ones, but square pots take up less space than round ones, so I do use them for cuttings and seedlings. But after three months there is a perfect clay-versus-terracotta race going on. Of the two identical cuttings in identical soil, the plastic plant is definitely winning. This may be because it is nearer the light, so I am going to bring both indoors, to the south-facing windowsill, where it is hotter than in the greenhouse, and then they will get equal amounts of light. Puzzling. I wish I was more scientific. William had the plant in terracotta and I kept the other. His has now doubled in size, mine is much smaller.

Every other Friday there is a tiny farmers' market outside the post office. Organic veg grown locally is impressive. Kalettes – which for some reason second daughter Ellie and I forgot to organise this year – and huge celeriac, not tiny knobbly ones like mine. They also have chard, beetroot, kale and winter salad. I have those too, but only endive, the mustardy leaves at the market, would be worth growing next winter. I ask if they are doing 'no dig' and the answer is, not quite, as they are taking on new weedy areas. Charles Dowding, who grows veg better than anyone, manages to clear patches with cardboard and compost. We tried this for the cut-flower patch in the churchyard, but the cardboard needed renewing and brambles were never quite vanquished, although they did at least become much easier to remove. At the allotment, lots of plots are covered in black plastic, weighted with tyres. Ellie, who is the best veg grower in the family by miles, doesn't veto this. I do because I think it suffocates the ground. And looks horrible. The cardboard route means you can add compost on top and keep the bugs happy. The plastic method apparently works better in summer when the sun helps to sterilise the soil as well as kill weeds, if sterilising is what you want, that is. I always mean to sow a crop of green manure over the winter, but it's too late now, so I leave any annual weeds, which I won't allow to flower, as this seems to me nearly as helpful as growing green manure.

Just written about snowdrops for *Country Life* and learned from talking to the expert that when you cross two species together, their progeny will be stronger, which means yellow snowdrops will be what galanthophiles call 'less miffy'. The yellow snowdrops are not actually yellow, they tend to have gold ovaries and sometimes a bit of gold on the inner petals. They can sell for over £1,000 on eBay. I have one yellow, G. 'Spindlestone Surprise', which does seem to be a cross of two species, and I am happy with that, although the Dryad snowdrops are probably more vigorous. They are miracles of plant engineering, but I love snowdrops in an ordinary garden way, especially the before-Christmas ones. G. 'Three Ships' is already out in the first week of December, and many Mrs Macnamaras are shooting green under the apple tree.

'Mrs Macnamara' was named for Dylan Thomas's mother-in-law, who kept an untidy house with two thousand books where Dylan and Caitlin occasionally lived. They would also stay in my village with John Davenport, a critic, writer and drinker who filled his house with musicians and artists in the first summer of the Second World War. The house is fifty paces from mine, tall, rather grand and a bit gloomy. Dylan Thomas had just published *Portrait of the Artist as a Young Dog* and, while living up the road, he and Davenport collaborated on a little-known work called *The Death of the King's Canary*, a fantasy about choosing a new Poet Laureate. It was something which Dylan Thomas had already started writing with a couple of other collaborators. 'It could be the best fun and would make us drinking money for the year,' he wrote to his Welsh friend Charles Fisher, but Fisher was called up for the war so Davenport became another co-author. Too scurrilous to be published at the time, it has some fine parodies of their contemporaries. How good is this of Eliot?

> Everything is the same. It only differs
> In the subjective mind which is the same
> Being or not-being, born, unborn
> A wind among leaves, deciduous or dead
> It does not matter where
> It does not matter.

What matters today is panicking about leaving plants for over a fortnight while I am away on Syros for Christmas, where Laura and Rupert spend half their life. Who will scare away the black cat from the blackbirds and robins who feed on the step in the courtyard, or under cover of the chair if it's raining? Who will fill up the feeder for the goldfinches and blue tits? Who will see that the *Verbascum roripifolium* seedlings on the high shelf in the greenhouse don't dry out? This delicate butterfly of a mullein is going to be my new favourite plant. Derry grew it last summer, so I bought two packets of seeds and sowed one, even though Derry, the great propagator, said wait until spring. But they are up, six tiny plants in cells, and I can't bear to lose them. The faithful daughters will come and Bob, the best of all neighbours, will open and shut the greenhouse. If he can't, Belle in the shop says she will do it. The logistics of leaving are alarming. I am rarely away from home for more than five days. In summer I leave notes about watering too many pots; in winter I am more likely to warn against overwatering, the real killer in a greenhouse at this time of year. Pelargoniums can tick over with nearly nothing. But the lemon tree needs a winter feed once a week, cuttings might need a few drops, dead leaves need removing and air flow is vital. My greenhouse has doors at either end and, in summer, the top vents open automatically. If it isn't actually freezing, the doors are kept open all day. If nobody is available to open and close, I leave the fan running to keep the air moving. Grown hard, plants are much more likely to resist disease and survive winter. I must remove any dead or browning leaves before I go, because that will keep trouble at bay. The plants on the windowsill at the front of the house must be carried to the greenhouse, or they may suffer. I usually only water them once a week in winter, but if the sun shines in through south-facing windows, they will need more checking than is likely to happen in my absence.

Syros apparently means 'rocky', which it is. Very. The vertical town of Ermoupolis is mostly neoclassical with many ruins. Cats slink down the paved streets where piles of coloured leaves and bougainvillea bracts like Christmas decorations have been swept to one side. Laura's three children, William and I are staying in a strawberry-pink mansion next door to Laura and Rupert's unfinished house, and my room has cross light, which I have always wanted, and a balcony for breakfast with a view of the sea and distant islands. But it's downstairs, upstairs on marble steps to Laura and Rupert and downstairs, downstairs, downstairs, downstairs to the market and harbour below.

On the way to the shops, we pass a beautiful tree that I have never seen before. *Melia azedarach*, the chinaberry tree, is a type of mahogany. Laura says it has pale lilac-scented blossom in spring; now it has porcelain cream berries and is a graceful shape. It is hardier than an olive and its seeds were once used for rosary beads. There is a church shop here where we plan to buy candles, and Greek-speaking Rupert can ask about rosaries.

Christmas decorations are going to be varied. Some glitter and three tiny hollies in pots have been bought from the garden centre for the table, placed between green glass jars for nightlights. The rest is foraged. A driftwood branch will be the tree. As yet, what hangs on the tree is under discussion. Laura favours snail shells, which she will paint pink with dabs of gold and then pierce a hole in the shell so that it can hang by a thread. Piercing is slow, fraught work, and everyone will be here in a couple of days. I favour brightly coloured sweets. The downstairs downstairs shops are full of them. For evergreenery, it's plenty of mastic, *Pistacia lentiscus* with berries, but Rupert says only red ones, which they aren't always. We have also picked some Aleppo pine, with cones, to drape over the kitchen cupboard, and seed heads and stems of sea holly, wild carrot and sea squill to arrange on the windowsill. Emily the florist has declared she will critique our efforts when she arrives, which has raised the ante a bit.

When Emily and I went together to the Garden Museum's winter flowers week, which was strictly seasonal and home-grown, we saw lots of dried flowers. The inner courtyard, which Dan Pearson designed, had a terrific stand of proper *Nandina domestica*, the sacred, or heavenly, bamboo, which isn't a bamboo but a kind of berberis. By 'proper' I mean not one of the dwarf hybrids like 'Firepower', which is mostly what garden centres sell. Evergreen with creamy flowers in spring, and scarlet berries now, it's easy to grow and a perfect courtyard plant in sun or shade, but it slightly prefers more acid soil than mine. In Japan, the sacred bamboo is planted as a symbol of welcome near the door. We looked long and hard at the different displays. Shane Connolly, the high priest of 'grown not flown', had an altar to nature, with wands of scented shrubs and pots of snowdrops and hellebores. Shane says, 'You can walk down any street in summer and smell roses, but scented shrubs in winter are gold dust.'

I'm hoping that the dried stems we have just picked on Syros will remind Emily of that day. In hot countries, or places with drier winters than ours, brownery is beautiful. Alice rings and we tell her how pleased we are with our efforts. She asks for a picture, so that she can copy what we have done in Bristol. But I doubt she will find anything like the plant ghosts growing here. Today she writes, 'Absolutely beautiful. We will compete. I think we will fetch green branches from Wales tomorrow, which already gives us the Druid edge.' It's bad enough having Emily arriving for the florist's critique without added competition from home.

It is the winter solstice, which in England is the time to celebrate the coming of longer days, but here the light is never absent. Pearly on clouded days and translucent blue today. This morning the grandchildren and William came in on the enormous ferry, which swirls to a halt at the last moment in front of us. After their luggage and shopping has been carried from the tiny car along the 'flat' route (only eleven steps down and twelve steps up), the older generation sets out for a beach to swim, the younger ones go down to the shops for a last-minute Christmas panic, and Ivo bathes from the steps below the town. Then it's lunch outside at three. Midwinter, this place is sheer heaven.

If I was at home on Boxing Day, the mummers would be performing in their Old Time Paper Boys role, all along the High Street. Crowds come up from Bath to see second-generation villagers dressed in raggy newspaper robes, acting out a strange drama of death and revival. There is a town crier with a bell, followed by Father Christmas, Little Man John, King William, Doctor Phoenix, Saucy Jack, Tenpenny Nit and Old Father Beelzebub. The play was performed regularly until around 1880, when it stopped. In 1930 the Reverend Alford was interested to hear his old gardener recite some verses which he remembered from his childhood. The vicar contacted his sister Violet, a folk historian, who pieced together the memories of other old villagers to create the play in its present form. Nothing was ever written down and, in the best oral tradition, parts are learned parrot fashion from past performers, which can make for a strangely wooden performance, but I will be sorry to miss it this year.

I have seen the sea-daffodil, *Pancratium maritimum*, growing on sandy beaches in Corfu, but here it flowers from a nest of pebbles on the beach where we swim on Boxing Day. Just seed heads now, and I am tempted to take a few black seeds home, but it needs very hot summers to flower. Last time I tried a bulb in a pot it did not survive, although now that our summers are so much hotter, perhaps it might succeed. Looking it up, I find:

Both sea-daffodils and narcissus pointed to the transformation of life, to the inescapability of death. While narcissus tended to be a reminder of untimely death in the middle of spring, sea-daffodils announced the imminence of autumn when the flora generally vanishes. The sea-daffodil, therefore, was also a plant associated with the vegetation goddess Persephone, who each year had to pass through all the phases of life, including the underworld.

A few sleepless nights ago, I listened to Natalie Haynes on the radio 'standing up for Demeter', Persephone's mother, who searched and mourned so long for her stolen daughter.

I miss flowers here. No *tazettas* out yet, which my sometime design partner Pip Morrison thought we might see, and Laura knows where narcissi grow, but when we look they are still only leaves. We find some crocus and Laura claims there are a few anemones around. It's such inhospitable terrain that it's hard to imagine anything flourishing. Occasionally there are green clefts where olives grow, but any trees are stunted and even the pistachio here is low on the ground. The point of the place is the light and the views across a silver sea to the islands, which float nearer some days and on others are barely there.

JANUARY
Snowdrop

Every email from England has complained of days of grey rain. Leaving Athens in sunshine, I already dread the lack of light, but during the last few days of our idyllic life on Syros, I began to long for green and the first signs of growing things at home. Travelling with William, we console ourselves by looking at how many more minutes of daylight there will be by the end of January, which is almost an extra hour.

Now I am home, it is mild outside, and birds are singing. The air is soft and smells of earth and flowers. One aconite is out, and several 'Rijnveld's Early Sensation' daffs are in bud. Primroses seem to be taking over the flower beds as well as the meadow. There is *Daphne bholua* in flower near the house and blossom on the prunus. Hellebores are appearing and masses of snowdrops are up, with 'Three Ships' already over and 'Mrs Mac' everywhere. The red pasque flower has mistaken Christmas for Easter and has four perfect flowers. That's many more plusses than I expected or deserved. On the downside, crocosmia stems have fallen over and tree peonies have a bedraggle of leaves, so those ought to be cleared. Seedlings are sprouting in the flower beds, so as well as wild, it must have been mild. I'd better leave them until

I can see what they are. Creeping buttercup has crept under a salvia and the perovskia, but the hori hori knife makes short work of that. Most of the prickly cat defences are in place, but one dahlia has had mulch dug away. After eighteen days of absence, while wind and rain apparently tore this country apart, my garden seems to have gone on happily growing. The seedling verbascums on the top shelf are fine and Alice has come a couple of times a week to check the plants in the greenhouse. She reports one squash on the front windowsill turned to mush and that the bean poles at the allotment have blown over.

We were lucky with the weather, Laura says. Windy and cold now on Syros, she reports, and that I must remember winter isn't always as benign out there. So, I remember January in Corfu twenty years ago, where I was creating not a garden so much as doing what Pip calls 'making a place more itself'.

Everybody says 'Going to Corfu? Oh, you are lucky.' 'Work,' I say, and they make a pull-the-other-one face. You can't get to Corfu in winter except via Athens with Olympic Airways, who are on their uppers. At Athens the local flights are all delayed 'due to operational difficulties'. Tim Hatton the architect and I wonder if this means that there is only one aeroplane for three destinations, which has to make a trip first of all to Kos, then back to Athens, out to Thessalonika, back to Athens again and last of all to Corfu. But we finally get going and as we start the descent it is raining and very bumpy. I think I hear the air hostess say we cannot land at Corfu due to weather disturbance. Tim has already explained that owing to the state of Olympic Airways' finances, only minimal petrol is carried. Panic stations. Tim tells me that what I heard was 'cannot hand out coffee' rather than 'cannot

JANUARY

land at Corfu', and we are down. The taxi driver who meets us assures us it is set to rain for a week.

Next day, three Albanians are on site with Irish potato shovels and iron wheelbarrows that weigh a ton. The earth is as sticky and heavy as cement. The Albanians speak no English, so they cannot tell us why the bulldozer has not arrived. The home team – Andy, Karen, Terry and Nico – have achieved everything on the last visit's list of tasks. The planted areas look better than I had hoped, although there is no sign of *tazetta* narcissi in the olive grove.

I set out plants for the courtyard. Honey-scented spurges, China roses, pomegranates and the odd almond, wishing that even this small area designated as a garden was being returned to the wild. It is icy cold, but blue sky, blue sea and huge views like the paintings of Edward Lear spread out in every direction. Olives as old as cold water spread their settled calm over the hill. The place is always more beautiful than anywhere I know; it makes me think that any change will be sacrilege. But the lack of a bulldozer is desperate. At the end of the road two men are tinkering with a dead JCB, which is anyway too big for the work which needs to be done. If the landscape is to be ready and planted by Easter, the grading of the ground around the tower must be finished on this trip. Thanassos, the king builder, arrives at 12.30 for a meeting that was scheduled for 11, in new and very clean shoes. 'No problem. Off corsss it will be finished,' he says. Tim is gloomy. Work on the tower is not completed. Work on the house has not begun. The bulldozer we need is in action somewhere else and we cannot have it. Somewhere else, Andy and Karen say, is new houses which must be finished by the start of the tourist season. Corfu Villas lends money to the locals to build or improve their houses so that they can be let to holidaymakers. This arrangement means no income for the owner for the first five years until the loan has been paid off; after that the locals grow rich on summer pickings. On other parts of the island the huge olive trees, which look like Arthur Rackham illustrations, are being felled for timber because the olives are not worth picking any more. Lorries come from Athens to collect the wood. One of the gardeners says, 'It snow tonight, upstairs,' pointing at the mountains, which are already dusted with icing sugar. Next morning, only two miles away on local hills, snow cloaks the olives. No bulldozer, no Albanians, and a wind so cold it is hard to remember what Corfu is like when you can barely move for the heat. Behind Butrint it is white too, and in the tower the shutters are forced apart by the wind. We light fires to test the chimneys in case they smoke and hope that tomorrow the bulldozer will come.

Back at the house it is spring-cleaning in the rooms, which are cleared in readiness for the builders, who are busy elsewhere. We keep the fire in all night, but our breath still clouds the air as we talk about whether a hedge will be as good as another length of wall. It won't of course, but when you are in a hurry, plants can fudge a lot of things. Tim's approach is direct and intelligent, and he listens. Sometimes with architects it feels like a competition for space, time and money, but not on this job.

Monday morning and so cold I wear a jersey on my head. Five Albanians in gym shoes and sweatshirts have arrived on scooters to move earth by hand. They get paid 80 drachmas a day. Greeks never do manual labour now that there is a plentiful supply of

JANUARY

Albanians. Vassily, who speaks English and Greek as well as his native Albanian, translates for me, and his team shifts a mountain. The bulldozer never materialises. At 3 p.m., freezing, we finish, and I go back to the house to eat lunch, which has been kept hot. Tim is having a meeting with Thanassos and Jimmy, the Greek factotum, with car park plans on the table. I say through clenched teeth, 'The Albanians were wonderful. They have almost finished the job. By hand. Because the bulldozer never came.' At my outburst, Tim looks more embarrassed than the Greeks.

Very cold at home today, in a sullen grey way. This is the time of year when it's hard to believe in summer. I try to remember having supper outside after a hot day, when flowers smell sweeter than ever and swifts dive and scream through the sky. They have nests in the village, but I wonder how long they will stay once every building is changed and modernised. There is a move to install swift boxes on tall buildings. This house is three storeys high, but bird buffs say they prefer the area near the church. I've moved the wood delivery and stacked all eight barrow-loads and removed a few dead stems of aquilegia, but I can't think of anything else useful to do outside. The ground is still too bony and wet.

The Met Office has been on full 'situation awareness' alert, with warnings of heavy snowfall, so I have given the mimosa tree in the courtyard a dress which looks like a shroud. Two thicknesses of fleece are tied to a tall stake in the walled corner behind the tree, and then draped over a semicircular iron border support. But now I look at the forecast again, it seems that southerners do not need to be as aware of the situation as Manchester and

Scotland. I love a mimosa for its bright yellow puffs of flowers, which dangle from silvery branches and smell of southern spring. But it is not a sensible choice outside London or Cornwall. People say climate change means we can grow Mediterranean plants, but it's the combination of wet and cold that kills, and the weather goes to such extremes these days. My tree did flower two years ago, before being cruelly halved last winter. I want to keep trying to grow mimosa for the memory of Rome in February, when florists sell bunches below the Spanish Steps. We stayed once in the flat above the Keats–Shelley museum. I wonder now why it wasn't more haunted. Keats died of consumption on the second floor at the age of twenty-five. We were on the third floor, and all night long tourists gathered to play guitars and sing beneath our windows until the water cannon swept them away at dawn. We hired bicycles and rode to Trastevere for lunch, or went to the market in the Campo de' Fiori to buy salads from old women with baskets of leaves on their laps. I suppose we did some sightseeing too, but what I remember is eating outside under a pergola dripping with wisteria. And the bunches of mimosa. That's what I remember, but it can't be true because wisteria and mimosa never coincide, so all the memories of many Rome visits must be scrambled.

Jane comes to garden. We start the morning by wondering what on earth we can do. I'd left some 'Antoinette' tulips to plant – Jane hasn't planted hers yet either – but that is going to be impossible in bone-hard ground. Cutting down dead crocosmia leaves and restraining the periwinkle in the passage that leads to Back Lane turns out to be doable, and we find snowdrops nosing through the gaps. Because the ground is so hard we can walk on the beds for a bit more tidying. Hellebores are all over Instagram but none are out here. Jane says hers aren't either and adds, scornfully, 'I'm sure those Instagram pictures are last year's.'

It's still cold this week; it was minus six last night, but the sun is out, and the prunus blossom hardly browned off and there is plenty of viburnum and Christmas box for the kitchen table and the lettuces I sowed at the weekend are UP. Is there anything more exciting than pinpricks of green? No broad beans or sweet peas, though, which were sown at the same time as the lettuces. I never autumn sow these, because the sweet peas would have to stay in the frame all winter and broad beans never make it in the allotment, where I know there are rats. Rats are here too, apparently. Martyn the ecologist, two doors down, keeps bantams – very pretty silver-spangled Hamburghs. Sometimes they fly over the wall to land in the courtyard, which one hen did today. I knock on Martyn's door, and he comes round to collect Atalanta – they all have Greek names. He says, 'She saw a rat and it frightened her,' and adds, 'I never clip their wings so they can escape.' It takes a long time for Atalanta to be coaxed out from under the *Melianthus*. 'If I had mealy worms she'd come like a shot,' Martyn says, and while we are waiting for Atalanta to decide if she likes grain enough to exit the *Melianthus*, he tells me he is going to be away on a project for a month. I ask where the bantams will be. Turns out they are going to stay with a friend in Bath, where there are probably masses of rats. I do miss having hens, but I'm away too much and the garden is too small to keep them here, because hens are great trashers of borders. I know Arthur Parkinson manages collaborative gardening with free-range hens, but he's unique.

Another revolutionary lecture at the Bath University gardeners'. This time it's Alys Fowler on polyculture. She is passionate and intense, emphasising everything with arm waving for far longer than the allotted hour, so Derry has to get up and issue an injunction to finish. If Peter Korn's advice was going to be hard to follow, Alys Fowler is going to make veg growing very difficult indeed. Polyculture means growing everything together, with no bare earth ever. No compost, no fertilisers, no growing in rows or blocks, no bought seed, no watering. It's messy, she admits, and the practice seems to involve here a lettuce, there a lettuce, here some kale, there a squash, here a parsnip, there some marigolds, a sort of 'Old MacDonald Had a Farm' approach, with everything allowed a shout. We learn that dahlias are delicious, and that broccoli grown in a sea of clover will fool the cabbage whites. A female butterfly can pick up the scent of a brassica from three miles, but when it comes to laying her eggs, she is too stupid to know the difference between cabbage and clover.

If I had more room and more time, I'd like to try some of this, but the hardest advice of all is to leave vegetables for an entire lifespan, allowing them to flower and then to seed and to cross-pollinate themselves lavishly. You grow on the seedlings to see which will be strongest and tastiest, allowing the chosen few to seed themselves again the following year. Repeat this process for several years, because, by selecting the best and tastiest plants each season, you end up with plants which have adapted to your particular site and soil, your very own seed mix. I do not have enough years left to get through the messy unproductive phase, so it's back to Sarah Raven and Real Seeds for forbidden hybrids.

JANUARY

I must get the seed order in this week. There are plenty of leftovers from last year in the box, but old seeds can be doubtful starters. I should have put them in tins, with silica to keep them dry, rather than in a wooden box in the kitchen, and in the damp potting shed. Derry is my choice for flower seeds if I want to be sure of getting exactly what it says on the packet. The Dixter form of the best marigold *Tagetes* 'Cinnabar' is taller than other commercial versions, and I think this is only reliably available from Derry or Dixter. I could of course have saved my own, and sometimes I do, but last year I didn't. Seed saving has to be done when seeds are ripe and dry. While things are flowering, they get deadheaded to keep the show going, but if I go away at the end of summer, I often return to find soaked brown flower heads which are never going to germinate.

Topiary is so lovely in winter. I miss the four yew columns on the raised terrace in our last garden, with the great cake stands of yew above and the hen at the top of the hellebore beds. When I think about them at around this time of year, I wonder if I should have planned more topiary here. The three box bushes separating the path from the greenhouse and the orchard were always meant to be giant bulges, ending in a collection of more spreading bushes to hide the bins by the back gate. When we came six years ago, I ordered the 'Rotundifolia' box, a fast-growing form with large leaves, making wide bushes which can be clipped into impressive shapes. But the nursery sent 'Handsworthii', which is a much more upright form. The bushes are a good height, but they will never bulge. I should have changed them, but I didn't, and now I wish I had some giant vegetable sculptures to see me through the dark days. *Buxus* 'Rotundifolia' is hard to find, but I have grown some cuttings taken from a friend's garden. Six tiny ones, which I think I will add on either side of the Handsworth ones to fatten them up. It will take ages.

The modern topiary gardens I most admire are the organic mounds at Great Dixter afloat in a flowery meadow, and the gardens made by Charlotte Molesworth, the artist, and Louise Dowding. Charlotte has peacocks and quirky shapes; Louise's coiled topiary looks like organic sculpture. Beautiful, but this place is tiny, and I want an airy ephemeral garden, with a chance to grow many more plants than evergreens. Restraint is what the landscape architect Nan Fairbrother called 'the beauty of one thing'. But I can't give up growing flowers. Lots and lots of flowers.

The garden here has plenty of scope for different kinds of plants. The courtyard outside the kitchen window is for trying to get away with tender things, a mimosa, a white banksia, two *chinensis* roses, *Melianthus* and lemon verbena, with pots grouped on the steps for bulbs first, followed by pelargoniums and salvias. From the courtyard, between my neighbour Bob's wall and the separate room opposite the kitchen, you walk under four arches, two covered in *Akebia quinata* and two in the rose 'Adélaïde d'Orléans'. I regret the akebia now, which has cream rather than purple flowers, but I thought the arches would be difficult to cover because any climber planted has to share a root run with the lonicera hedge. Akebias are notoriously unfussy and vigorous growers, but the rose has been fine, so Adelaide could have covered all the arches. Emerging from the corridor, the size of the opened-out garden comes as a surprise. Behind the outdoor building there is a shady paved area with primulas, snowdrops and ferns. The scrap of lawn, with washing line, occupies half of the first third of the plot. The other half of this area, as far as the greenhouse, is made up of three flower beds divided by a straight gravel path and a curling grass one. This is the gardeny area, with many old-fashioned plants which modern gardeners avoid because they need feeding and watering and staking. Beyond the beds, the last third of the garden is all meadow, with three apple trees, a pear, a plum and a damson at the far end.

Last year I thought I should turn the patch of lawn over to gravel, for the chance to do a bit of modern gardening, with grasses and sparse plants which need no water. I thought about it a lot and consulted Jonny Bruce, who cares for Prospect Cottage, Derek Jarman's garden of flotsam and jetsam at Dungeness. For several months I got excited by the idea of doing something different, until I realised that this was totally out of keeping with the rest of the garden, and that the space made by the small lawn provides a pause after the courtyard and before the floweriness of the beds and the orchard meadow. I like a garden that flows, not one of episodes, and even in this tiny plot I want the walk round the garden to feel like a continuous experience.

The temptation of growing different kinds of plants in a small space is constant. I even want those which need acid soil. I can resist most which would hate the conditions here, but there are some winter flowers which I want so badly that I experiment with putting them in pots. *Edgeworthia chrysantha* has tiny hanging parasols which open yellow and scented. It's a weird woodlander, but any winter flowers are welcome and it's nearly out now. Most welcome of all is *Hamamelis × intermedia* 'Pallida', the pale-yellow witch hazel, which is flowering already, in a big pot in shade. This year I planted seven for a client on damp acid soil.

A grove of witch hazels puffing heady scent in January: it's hard to think of anything much lovelier on a winter morning. Later on, there will be more scent from *Rhododendron* 'Fragrantissimum', in another pot tucked into the angle of the wall next to the kitchen. Technically not quite hardy, this has survived all our winters here. I want to try *Camellia sasanqua*, which flowers for Christmas, but that will need another large pot in semi-shade, and I must leave room for spring and summer plants. All the winter acid lovers are permanent and much too heavy to move when their season is over, so I can't have everything I want.

The seasonal competition for space also happens in the flower beds. In the early years the lemon-yellow *Coronilla valentina* subsp. *glauca* 'Citrina' was queen of the courtyard, but now that the white banksia rose has got going and is flinging branches everywhere, the courtyard is much shadier. Gardens never stay put; as one thing grows, another declines. Sometimes the thug is allowed to win and the planting around it has to change. Sometimes it's the moment for a drastic cut-down. But for the time being the banksia can dominate. I could always put the coronilla in another pot, I suppose.

FEBRUARY
Algerian iris

Ann Christopher, the sculptor who lives in the village, invites me to see her snowdrops. It was her husband Ken's collection, so she is not sure of all the names, but there is a list with photographs indoors, which she plans to update. She has a huge clump of 'Wasp' among some ferns. 'Wasp' is a snowdrop which I thought I didn't like. Too spidery, too creepy, I always thought, but today I see the point. Its petals are long thin threads, like wings. It's Ann's favourite. Sculptors see shapes, and when I look closer, 'Wasp' is intriguing, and in a massed clump the flowers are airy and delicate. She also has a patch of 'yellow' snowdrops 'Madeleine'. Galanthophiles get very excited about any aberration from the norm. My little 'yellow' one called 'Spindlestone Surprise' has not increased much. Ann's 'Madeleine' looks a better bet.

The snowdrop collection is interesting, but it's indoors that's the thrill. There is a plaster Frink head and a Pither stove and a table with found objects: stones, sticks and seeds, all arranged in little clusters. On the floor there are more stones, not a Kettle's Yard coil of perfect round pebbles, but two rows of three lumpy coral fossils. Five are single stones and in the sixth place there is a stack of three. The bottom stone is flat black, above it, slightly smaller, is a round black stone, and the top, smaller again, is the mottled coral kind. I don't ask, but I wonder why. Was the sixth stone too small to stand alone? Or was it just too dull to make the double rows match? Sculptors see mass and void first. I see colour. The stones I collected on Syros beaches at Christmas were chosen for their different shades, and they live on a pink lustre plate, while misty blue glass shards from Brittany lie on a majolica pattern.

Tomorrow is snowdrop study day at Colesbourne, and although I said I wouldn't bring a projector, now I worry that there will not be the right connectors on site. I ring Currys in Chippenham to ask what I need to connect an iPad to a projector. They will not be open until ten and if I want to hear John Grimshaw, who used to run the garden at Colesbourne, I won't make it. I should have thought of this sooner, and by the time I start emailing about connectors, the office is closed and my emails go unanswered. In belt-and-braces mode, I send the slides via MailBigFile from bed. Next morning John Grimshaw emails to say everything is on his computer now.

Feeling thoroughly unprofessional, I set off in rain to arrive in good time to hear John speak forcefully and delightfully about the revival of the garden and the snowdrop collection at Colesbourne. It's a marvellous place, not just for snowdrops but also for the drifts of naturalised *Cyclamen coum* and *Crocus tommasinianus*. It's impossible to buy the true silvery species of this best of all crocuses, and I really miss having them here. When we moved, I ordered a thousand which turned out to be 'Whitwell Purple'. This is too large, not my favourite colour for spring and does not self-seed, but that's what bulb nurseries sell. Being sent the wrong bulbs is exasperating and it happens much too often. Since the dud order, I have scrounged a few true Tommies from friends with gardens where they grow and, every year, I dig out more of the strong purples. John described collecting seed from the crocus at Colesbourne and broadcasting them. And waiting. There is a patch on a bank in a village near here. I must remember to look for seed in May. A better bet may be the rumour that Troy at Sissinghurst has been heard saying there are too many Tommies in the Lime Walk. Can there ever be too many? If this is true then I want to be in the queue for bulbs that are being discarded.

I came away from Colesbourne thinking about abundance. Here, at this time of year, the garden seems skimpy, partly because of its size but also because of its age. At Colesbourne, snowdrops were emerging from duvets of *Cyclamen hederifolium* leaves. Hellebores were crowding each other among ferns and other snowdrops. *Narcissus pallidiflorus* was seeding between crocus, crocus was seeding everywhere. No primroses, though, which do seed here, and later there will be masses of cowslips. John told us that he disliked growing snowdrops in flower beds with later plants, because they are bound to get disturbed. True, but small gardeners have less choice, so my snowdrops are clustered under the apple tree and near the wall, where I hope they won't be bothered by inserting a half-hardy for summer. But it's dry around the tree and next to the wall, so the advice I took home was that all snowdrops are greedy feeders and need water in dry spells until they have completely died down. We had questions after my talk, and the old chestnut 'Do you move snowdrops in the green, or do you wait until June?' came up. John is adamant that if you lift bulbs while they are still growing, roots get torn, and the bulbs take a long time to recover. This is a counsel of perfection from the Galanthus Guru. Most of us are too busy to find snowdrops in summer and move them, so in the green it often has to be.

At last, the old shop opposite me has a roof again. Just over a year ago, I was on site in a garden near Bibury when a friend texted to say, 'There's a terrible fire opposite your house, you won't get into the High Street. It's blocked by fire engines, it's all very frightening.' I got home about four and left the car in the market square to walk the rest of the way through police, fire engines and plenty of people. Central Stores was the hub of village life, Sally and Rachel, the sister owners, would be devastated, but even worse, the fire had spread to their neighbours on either side, who were now homeless. As I made my way home, I was stopped a few times to be told that, at one point, there were sixteen fire engines in the village.

As well as the completely burnt out and now unusable shop opposite, the sisters owned a second shop on my side of the street, an amazing cavern with mahogany shelves and drawers, once used as a set for the *Cranford* television series. This was Bodman's Country Store, owned by five generations of the same family for over a century, until it was bought in 1980 by the present owners' parents. One half of Bodman's was a draper's and the other a grocer's. There are still drawers with handwritten copperplate labels saying 'children's handkerchiefs'. Since we arrived in the village six years ago, this second shop has stocked newspapers, dog food, birdseed and some gifts. The sisters had been trying to sell it, but because the interior is listed, it was hard for anybody to imagine how to run a modern business there, so it remained unsold.

As I passed the entrance, two doors from me, I could see huge activity in the paper shop. There were people on ladders with buckets and brushes, people carrying boxes, some crying, others bringing cake. It looked as if half the village had turned out to help clean shelves, to sweep and mop and get the old stores ready to become our new shop. 'We'll be open tomorrow for milk and bread,' Rachel said. And they were. I volunteered to bring more hot water and cloths from my kitchen and joined the scrubbing brigade. Then Sally, Rachel, the chief fireman and their insurance man had a meeting in our kitchen as this was the nearest quiet place. A boy of about fourteen stopped to say, 'Is there anything I can do to help?' He fetched a few buckets of hot water and when I remembered I had left my telephone in the car in the marketplace, I said, 'If I give you the key, can you unlock the car and find my phone and bring it back to me?' wondering, as he ran off, if that was entirely sensible. It was; he was back almost as soon as he went.

It's hard to remember how bleak those first days were, but gradually the new old shop has been spectacularly transformed. Between the wide wooden counters there is much more room for gossip central than there was in the old shop. The shelves are stocked with provisions, the drawers are filled with pens and tacks and Sellotape and other useful things. There is a cheese counter, wines, washing products, vegetables and fruit on the pavement under black awnings, and two discreet freezers as well as more tempting breads, homemade cakes and biscuits than is good for anyone. When friends come to the house, I always take them to look at what is now the Country Stores. Rachel says, 'People come up from Bath just to see us.'

This month I have another very old birthday. Perceptions of age come and go, advancing and retreating, in and out of view like those Greek islands. It all depends on the internal weather. Now sixteen years older than Penny Hobhouse was when I went to have lunch with her, I remember thinking at the time, shouldn't she calm down? Be old? Retire?

I am late for lunch with the garden designer and writer, Penny Hobhouse. Very late. A puncture two miles from home and I, an unreconstructed woman, cannot change the wheel. The jack would not jack, and the nuts stayed stuck, but summoned by the mobile, Charlie comes to the rescue and gives me his car, so that I can depart forty-five minutes late. At Mere, I ring Penny to say where I am and why. 'What an extraordinary route,' she says, and tells me I will be another hour, but that it doesn't matter as I can overlap with her next engagement. When I do arrive, hands black, clothes crumpled, all apologies and explanations, she says, 'I thought cars didn't have punctures these days,' and offers me a bath.

Her coach house is cool and full of beautiful things. Serious works on global art, not coffee table stuff on gardens, lie on tables. Her own books are stacked in neat piles on the stairs. We have lunch on a dark cloth like those carpeted tables in Dutch paintings. Potatoes and salad from the garden, smoked salmon and strawberries. Penny – well-cut hair, beige trousers, intelligent face – radiates authority and intellectual rigour. She is a true professional, an Oxford woman who has found herself through work. Her formidable skills at organising her time are legendary.

I am exhausted by the thought of her frequent trips to the States, her two books a year and her tidy house. Where do the grandchildren pictured over the bath (which I did not have) fit into her life? 'Modern technology is amazing,' she says. 'Digital cameras can email pictures of work in progress, so that you can see what is happening every day.' I say, 'But will it show the colour and texture of the pointing on a wall?' She answers that she trusts her contractor and her partner on the other side of the Pond. Is delegation part of being a true professional? We talk about other designers. Of one she says, 'He makes the most elementary mistakes, paths leading nowhere. *Terrible* taste.' The terrible taste topic is one I used to enjoy, but since my espousal of pluralism – gnomes are OK, you can have what you like as long as it suits you and your place – I am torn by the taste question. *Gardens Illustrated*, of which Penny is an associate editor, has crystallised the notion that the perfect garden is something to aspire to. I believe good gardens are a fusion of personality and place. All that matters is that you do it 'My Way'. The Frank Sinatra style of gardening can lead to tack and tastelessness, but I like the results better than those where people try to copy the latest look. Nor do I want the static perfection of a magazine photograph in my own garden; I like plants which self-seed or flop onto paths, so that you feel the place is getting on with growing in its own wild way.

Penny's garden is enclosed by the walls of the old kitchen garden of Bettiscombe House. From her modern garden room, you can see it all through plate-glass windows. This room is terrific, an airy space overlooking the contemporary, intellectual, rather masculine sort of planting which she has arranged outside. 'I like threatening plants, spiky artichokes, thistles, big things,' she explains as we gaze through the clean glass. Evergreen shrubs now punctuate the grasses and perennials. As their owner ages, they will spread, so that the work of maintenance should contract as her skills decline. This rational approach is obviously sensible, but my own inclination is to keep going in a sea of pleasing decay. With any luck the 'sans eyes, sans everything' phase will make it seem as though the familiar flowers are still there. Evergreen shrubs are too solid for me, but her *Viburnum × hillieri* is a very desirable plant. In an old laundry copper, similar to mine, she has gone for the beauty of one thing: blue grass that matches the greeny blue of the metal. It is a much tidier option than the *Cerinthe*, heliotrope, marguerite, geranium, *Francoa* and co. which currently fill my version at home. The whole place is tidy, but catmint flops on the path. I wonder, too, how much this garden at Bettiscombe changes; the permanent underpinning of shrubs must make it quite predictable and much less seasonal than other gardens which major in perennials. But I do admire the overall effect, which, like its owner, has an organised, professional look. No off days here, no rain-flattened soggy petals ever, I suspect. This is not a place where you might catch a bad flower day.

    If it seems to the visitor too good to be true, the owner notices all its imperfections. As we walk up and down narrow paths between the plants, Penny utters despairing cries and dashes indoors to fetch the secateurs. 'I can't come outside without seeing things that need to be done,' she says, apologising and disappearing again to fetch some water for something in a pot that I do not recognise and cannot quite bear to admit that I do not know. I follow her to the water tank where galvanised cans wait on the gravel below. Exclaiming, 'The cans are not even full,' she dunks them and puts them down again. I know good gardeners do this to warm the water, but most of my watering is done at high speed from a hose with a lance on the end. Someone should conduct a few controlled experiments to see whether plants really mind having cold water thrown all over them.

'Flowers are less important to me now,' Penny says as we wander back to the front of the building where a sheet of water, a rectangle of silver, lies under the beautiful Dorset hills. We walk down a mown path, meadow grass on either side, to her wildlife area. The pond is for toads, but the coping is slippery stainless steel. From a distance it looks wonderful, a continuation of the water, but as we approach, I worry about the toads on that cold and shiny stuff.

All this has been accomplished in just over an hour. If I leave immediately, I will not need to overlap with the publisher. Penny's day is heavily structured. Noon for the luncher, 2.30 p.m. for the publisher, 4.30 p.m. a tour of local wild flowers. Tomorrow, she drives to Wisley (six hours in the car) for the day to supervise the building of her country garden for the RHS. Over lunch she had confided that she was doing less. What would be more?

Will I ever be as in control of my life as Penny seemed to be that day? Saturday, I caught an early train down after two days in London of seeing friends and Impressionist drawings, a slightly too luscious French film about cooking, Holbein drawings and a thrilling Pina Bausch ballet. I stayed in William's new flat for the first time, and we breakfasted together, before he left for five months playing Putin on Broadway in New York, and I set out for Potato Day in the village, followed by giving lunch to a friend who I rarely see. Queuing for 'Belle de Fontenoy' – only three left, so I had to make do with 'Charlotte' – flapping about getting home to find unopened letters, dead flowers on the kitchen table, buying cheese at the shop. I was short with an allotment neighbour who wanted to stop and chat. 'I've been away,' I said, 'and only just got back. Got to make lunch for a friend.' Mairi looked puzzled. 'You *are* cramming it in,' she said. Should I calm down? Be old? Retire? The pace of life is slower and friendlier here. I love it when I live it, but the change from action to reflection is sometimes difficult. The garden is the best place to wind down, but in the afternoon it rains too much for anything but greenhouse tasks.

The hellebores are disappointing this year. Late to start and the buds of 'Green cups', my favourite, are only just nosing through the ground. I used to like the dark-flowered forms, even an almost black. But at this time of year, when days are short, pale colours stand out against dark earth. But 'Anna's Red' in a pot looks so good that I think of dividing it: some for the pot and some for the flower bed. On Instagram, Edward Flint, ex-Dixter, says, 'Division of orientalis hybrids shouldn't be too frequent and isn't suitable for all hellebores. Those such as "Anna's Red" will die.' When I question this, he replies that it's a tissue culture plant, not strong and too woody.

The wealth of freely given knowledge on Instagram often surprises me. On garden.johns you can see and hear John Massey, perhaps our greatest nurseryman, talking about the plants he grows. This week it's shocking-pink *Cyclamen coum* that's painting the ground all over his garden. But he does say that these earliest of spring flowers grow naturally in damp places. Puzzling. In our last garden they seeded in the dry south-facing bank at the end of the lawn and seemed happy there. If you listen to *Gardeners' Question Time* on the radio, panellists often disagree about the best way to grow a plant. That's what I like about gardening: the hap and hazard of it, the chance to do things differently when something fails.

This rain is relentless. It should have been a gardening day with Jane, but we decided to put it off until next week. If I'm at home I can't bear not spending time in the garden. A couple of hours outside is enough, but today it's non-stop grey monsoon, so I have brought a pot of beetroot seedlings indoors to prick out into the new propagation tray on the kitchen table. The potting shed is too dark and cold and the table outside where I normally work with seedlings is very low and often brings on a bad back, so I'm pleased to find sitting at the kitchen table with the plants on a tin tray is the perfect solution to rainy-day gardening. The 'no dig' Dowding method is not really suitable for direct sowing and anyway it's too early. I should of course have sown all the seeds in these cells at the start, but I only had the trays with sixty spaces, which I find too large for my needs. Sowing different crops in the same tray never quite works, because some seeds appear sooner than others and need to come out of the propagator, while others need a longer turn in the warm and under cover. The new ones which arrived yesterday have thirty spaces, which is going to be much easier to manage. The trays are rigid, with large drainage holes, so pushing the little plants out with a finger or a pencil is far better than struggling with the plastic garden centre numbers.

Kirsty Knight Bruce came to stay. Having been a painter, now a designer, she has the best eye. She thinks I am growing too many rarities, and when I look, she is right: this garden is far from burgeoning. *Prunus mume*, with shocking-pink flowers, isn't exactly covered with blossom. The mimosa is alive, just, but looking peaky. I've banned daffs in the orchard meadow because I want rarer bulbs like anemones and tulips, but when we go to Catherine's for lunch, she has daffodils and forsythia on the table and airy bushes of *Osmanthus* with white flowers near the front door. I never thought of *Osmanthus* as airy. It's useful as a clipped bush, but growing exuberantly, with space between the branches, it looks so fresh. Catherine's garden has the suggestion of spring while mine is still dragged down by winter. So today I resolve to grow what is happy, to let cowslips and primroses seed everywhere rather than keeping hopeful spaces for the *Anemone hortensis* which can't have enjoyed day after soaking day this year. Right plant for the right place is such an obvious maxim, but it's hard to resist the challenge of growing something unusual and difficult.

MARCH
Spurge laurel

Acid green signals the beginning of spring. Beacons of *Euphorbia characias* subsp. *wulfenii* or lime-bright hellebores are the colours in Paul Nash's Landscape of the Summer Solstice. While I don't think I want those colours so much in the summer garden, the piercing shout of a green that is almost yellow is a tonic after drab winter days. In my childhood there were fields of mustard, not rape, growing on the Berkshire Downs where my sister and I were at boarding school. I was seven and my sister was only just six, and when we went home for the weekend we laughed when we saw the bright patches from the car.

Orchard House in Blewbury belonged to Lady Forbes, whose daughters had grown up and married, and the school was run by their governess, Miss Kaye, after her charges had left. There were never more than seven of us, and the house had a beautiful garden. But I was homesick and came to see the garden as a refuge from school.

Do I dare, do I dare, run round the garden before the grown-ups wake? There are rules at my school and getting out of bed before we are called is not allowed. Going outside without permission is not allowed either. The bolt on the side door is heavy and it makes a bit of a noise. When I do get out, the grass under my toes feels cold and wet, and the gravel paths are prickly. I run as fast as I can until I get to the kitchen garden, where there is nobody about. Next to the path on both sides there are long green pillows of leaves like grass with raggy white flowers. If I pick one, even my sister will know I have done the dare. I should keep running, but I want to be out here a bit longer. I want to see the garden just growing by itself, waiting for the sun to be properly out. The flower I pick smells like the soap that comes in pretty wrappers at home. But it is cold and a bit scary and I run again, so that I get back to the warm indoors. Icy grass, and then soft carpets. When I get back into bed, a few blades of green stick to my feet. 'I could have done that. Anyone could,' my sister says, but she doesn't have the flower, and she didn't see the quiet garden lying there with no one around before the summer day begins.

In the garden that is usually out of bounds we are dancing the Valse des Fleurs after tea. It is a birthday treat for my mother. There is a tree with a bending branch nearly down to the ground. 'Don't climb on it, or you will fall,' but we can wait under it for the music to start. Miss Kaye winds up the gramophone. At the beginning birds sing, then we have to step slowly, and then we must jump and twirl and twirl and twirl, all round the lily pond. We dance with no shoes and the dresses are slippery. Mine is pale orange, but I wanted mauve. The pond is dark as a bar of chocolate, with round lily leaves and sometimes a toad on the edge. I know about the toad, because I come here sometimes on my own, only I don't quite go into the garden. I crawl under the hedge that smells poisonous and has red berries that the birds eat, so they can't be as poisonous as everyone says. When I go back after the dancing, I try to remember the music and the twirling, but it isn't the same and I can't stand up in the hedge, so I can't remember how lovely it all felt. The garden is very quiet without the music. I think gardens are just themselves when no one is there. They get on with growing and sometimes petals fall off the roses, but nobody sees.

We know now that gardens are healing places. On the wireless I hear Hanif Kureishi talking about how much the Horatio garden at his hospital helped him to come to terms with his disability, and Sue Stuart-Smith's marvellous book, *The Well Gardened Mind*, explains and endorses that. An hour of pottering among plants is still the most restoring and comforting thing I can do. It sounds like it always was. Antonia sent this from Victoria Glendinning's biography of Trollope, where in *Castle Richmond* he writes,

> Let all those who have houses and the adjuncts of houses think how considerable a part of their life's pleasures consists in their interest in the things around them. When will the sea kale be fit to cut, and when will the crocuses come up? Will the violets be sweeter than ever? And the geranium cuttings, are they thriving? We have dug and manured, and sown, and we look forward to the reaping . . . All men love these things, more or less, even though they know it not. And women love them even more than men.

And now, almost two centuries on from Trollope, at last the younger generation of today have pronounced that gardening is cool.

I still worry about this fact from the Horticultural Trades Industry, which states that 'approximately forty-three million people in the UK have access to a private garden. And sixty-two per cent of those with private gardens use them to grow plants, trees and flowers. That's approximately twenty-six million people.' So, what I want to know is, what are the remaining 17 million people doing with their plots? It bothers me that, for so many, a garden represents a chore, a sterile space where lawns are mown and leaves are blown in a round of noisy outdoor housework. Rather than perennials, expensive bedding plants are shoved into borders and tubs in spring, and pulled out and discarded in autumn. Shrubs and hedges are savagely clipped, so that the whole place feels punished rather than loved. I know all this because Ellie has been helping a friend who does contract maintenance in Bristol, and she says that these are gardens which have usually been laid out by designers.

The second day of the month and it's snowing. Unwelcome weather, which yesterday the forecasters did not forecast. At seven, from my window it looks pretty, but dismaying, just when spring seemed to be on the way. By nine, when I go to the shop for bread, it's already deep slush. No bread yet, but an hour later Mary knocks on the back door with the Bertinet loaf, even though I said I would collect it.

I must sweep the flagged passage under the arches before frost follows and the route to the greenhouse becomes a skating rink. Wet paving stones are dicey enough, but frozen ones are lethal. Clients with York stone terraces get their gardeners to power wash them, but it's a waste of water and the run-off is unhelpful to most plants, so I try to discourage it. Sometimes I throw sand or horticultural grit on the passage, but mostly I creep carefully and if it's really bad I keep a hand on the wall.

When we bought the house, the little courtyard was also paved with York stone. We changed this, but not the passage, because looking out at the courtyard all day long from the huge kitchen window stone is gloomy. When it is wet, the stone turns black, while gravel reflects the light and gravel drains. Best of all, you can grow things in gravel, so the dream of a gravel garden isn't quite over. Last autumn, I decided that the big oak table and four beautiful chairs were on their last legs. Literally. They were filling the middle of the courtyard, but these days eating outside is rare. At our last house there was room to store the furniture undercover in winter, here there just isn't, and everything is falling apart in plain sight. I've bought a smaller, not very nice, slatted table and a couple of chairs and tucked them into the corner behind the *Edgeworthia* in its pot. This is enough for a cup of tea, or coffee with a daughter or a friend, and it leaves room to plant the gravel. Two bluest viper's bugloss have already gone in, and the special *Verbascum roripifolium*, which I have been nursing through the winter in the greenhouse, should be happy there. It might even be a place for the *Anemone hortensis* grown from seed sent by kind John Morley and for scented *Gladiolus tristis* from South Africa, which can cope with winter wet but likes to be very dry in summer dormancy. People will probably tread on some of the plants and whenever they do, I will scream, but it's worth a try.

The daffodils I like best are tiny or old-fashioned. 'Bath's Flame', *moschatus* and 'White Lady' are tucked into corners where their dying leaves will not be too obvious. That's the problem with daffs: their leaves are large, and they need six to eight weeks after flowering before they can be cut down. In our last garden we had lots of *N. pseudonarcissus*, the wild daffodils in Herrick's and Wordsworth's poetry. They seeded everywhere, but when we moved, I wanted room for other bulbs in the little meadow. Jonquils with narrow grassy leaves (*juncus* is Latin for 'rush') are easier neighbours for anemones than most daffs. A new favourite is *N. cordubensis*, from Spain. Its bright yellow flowers are deliciously scented and it is early to rise. More reliable is 'Kokopelli', which seems as happy here as it is in a south of France garden where Pip and I have been working for several years. I've been less successful with *N. jonquilla*, perhaps because I put them at the shady end under the apple trees. *N.* 'Segovia' is a miniature white small-cupped daff which looks good in short grass and lovely among the *Anemone blandas* as they begin to flower. Latest of all is *N. poeticus* 'Recurvus', the old pheasant's eye, with its small flowers and the best scent. Nurseries offer 'Actaea' instead, which is bigger and earlier, but pheasant's eye is the one I want.

In the garden where I grew up, designed by Percy Cane, there was a daffodil walk, with banks covered in every shade and type of daff. My mother gathered armfuls that smelled of nothing much outside, but, in the flower room, where she arranged Constance Spry vases, the scent was yellow and green. 'Doing the flowers' was her thing. Not mugs and jugs on the kitchen table, nor winter flowers, ever. She thought snowdrops were unlucky indoors and once, when I showed her the lily of the valley scented sprays of mahonia in January, she could not see the point. At home, I avoided the parts of the garden where grown-ups walked and talked, but I loved the banks of the river where primroses grew.

A sunken area of irregular paving stones was another favourite place. Limestone-patched earth was once traditional around vernacular buildings. But clients today are suspicious of 'crazy paving', which this was not, because the stone is never jigsaw cut but just large pieces fitted gently together so that plants can grow in the gaps. In March, there were grape hyacinths so scented that after a bit you had to stop smelling them. There were scillas of piercing blue and pools of purple aubretia, followed by doll-sized rock roses. At the time I never knew the names of even these ordinary plants, but I loved the way they grew and often went to see them. Now that I remember what they looked like, I can give them their names. Much later I saw a paved garden made by Alvilde Lees-Milne at her house in Badminton village, which reminded me of that early place, and it is something that I have tried as a designer to suggest to clients. Old shibboleths die hard. Dahlias, orange, crazy paving – 'Aren't they naff?' the clients ask. Dahlias are now OK, but in the eighties they certainly weren't.

'No plant is naff if you like it and want it,' I say to the clients. Their choices are not always mine, but it is interesting trying to see a place through its owner's eyes. The genius of the place partly comes from what it feels like, its atmosphere and how it connects to the landscape, but those who live there have rights too. Plants are changeable, like cushions. They matter much less than space and the permanent shapes of trees, or hedges, or clipped evergreens. But at *Photinia* 'Red Robin', I probably draw the line. It seems to be everywhere these days, and its solid perpetual red and green leaves dismay me. What I love about plants is how much they change, how ephemeral they are, but the perma-coloured *Photinia* is such a fixture.

Working with Pip Morrison, we always start by looking – it is a kind of worship, a way of understanding the feel of a place. Eyes are any gardener's best tools. I use a basilisk stare to spot when something is out of sync with the whole composition here. Gardens alter with time and when you move through them, so you need to look at every season and from every corner. When I was on the National Trust Gardens Panel with Kim Wilkie, he used to run from one place to another to see it from another angle. A garden is a picture made up of tiny fragments, which you never see all at once. Like painters, gardeners select, discard and rearrange. Like them, the thing we make needs vision and patience and skill. We bring out what lies under the surface. Pip says, 'We make a place more itself.' For me, it's about making somewhere to be rather than somewhere to look at. When we meet clients, we get them to think about what they want the garden for. And instead of making lists of plants, I ask them for lists of words to describe how they would like the garden to feel.

The best compliment I ever had from any visitor to our last garden was when I found two girls whispering in a corner under an apple tree. When I asked them why, they answered, 'We don't want to break the spell.' And I still wonder what it is that works the magic, the spell. There is no formula and everywhere is different. Slowing people down is one way to get them to stop and stare. If you study garden design, you probably get taught about making quite wide paths leading to a vista. But the gardens I really like, the linger-in-the-head otherworldly ones, have narrow paths and sometimes things that make you duck and weave. You can't walk fast, and you can't see what is ahead. In a small garden like mine, the curling grass path to the orchard only reveals the meadow at the last minute. The first meander before breakfast and the last in the evening can still make me feel I am somewhere special.

I came late to growing produce, because vegetable growing was always Charlie's province. In the old garden I kept hens and used to let them range freely in the kitchen garden, which he tolerated – just. When we moved and took on an allotment, I began to muscle in on Charlie's space, only to discover that he wanted to grow many more potatoes than I thought useful, and to put rows of brassicas where I wanted salads or flowers. He was resistant to 'no dig' because he believed double digging was vital and he actually enjoyed the spadework. So our allotment was disputed territory, but it is the place where I miss him most.

I haven't been down for weeks and when I get there, I find the rhubarb is bursting out of the forcers, so that the lids have been tipped on the ground. I have to cut the leaves off so that I can lift the pots above the stems. There is so much rhubarb I will have to freeze it. I offer some to the mother of three boys who seem to be making a den in the covetable Keder house on the next plot. She looks doubtful and says, 'You can make gin out of it, can't you?' Their rhubarb was a bit of a failure last year, she adds. The broad beans which Ellie planted for me last week while I had flu have been flattened by snow falling on the fleece which covers them. In the greenhouse I have lettuce, beetroot, sweet peas and spinach to be planted out. I must definitely make time to get down to the allotment next weekend, although it is still wetter than I've ever seen it. I need to find some planks to work off, a trick I learned at Dixter, which is useful for times when the ground is too wet to tread.

At teatime I walk up the street to see how Sheila, ninety-two, is after her hip operation three weeks ago. On the way, Simon is giving the holly tree lollipops outside their door a drastic prune. His wife, who is watching, looks alarmed. I say, 'Probably fine,' as I pass, and hope I am right. Sheila is unattended. She makes a cup of tea and the telephone rings. Someone at her book club last week had said she was planning to put an armchair on the street, but that it might be a help in Sheila's recovery phase. Can she bring it round now? Sheila is one of the most visual people I know. 'It's a pink chair,' she says, looking mildly dismayed. When the couple arrive with it, Sheila is grateful but firm. 'When I am properly mobile,' she says, 'I will put it back on the street, so that someone else can enjoy it.' On sunny days people leave books, toys and surplus fruit outside their doors; now it seems chairs too. Facebook is another exchange-and-mart gathering of recommendations for plumbers, painters, outfits for World Book Day and sundry offers. It's a useful grapevine in a very connected village.

The first proper spring day with Jane after flu and days of finding fault with the garden. Jane is positive. Too many primroses instead of crocus, be grateful. The late snowdrop 'Peg Sharples' lights up the back of the narrow bed, and after lunch the *blanda* anemones open in the sun. There are tiny peaches on the 'Peregrine' peach tree and a seedling hellebore turns out to have a dark anemone centre. There are infant *Smyrnium perfoliatum* everywhere, so a few are moved under the big apple tree. They take three years to flower from seed, but it's worth the wait for their singing lime-green presence in April. Once established they need a bit of subduing, but perhaps, like the primroses, more is less. Less fussing about picky plants who don't seed happily around. Honesty and the little round wine-red *Allium sphaerocephalum* come up like cress and must be curbed, and I'm looking forward to the pink *Cardamine quinquefolia* running all over the corner near the pink magnolia. It likes shade and is such an early spring presence that I wonder more people don't know it.

Eco book group again yesterday. Gardening is currently confusing. Just when I have become a zealot for 'no dig', the hardcore environmentalists say, 'But what about all the bare earth between crops?' The answer, it seems, is covered ground at all times, more Alys Fowler than Charles Dowding. Rows are out. Permaculture and regenerative gardening, which mean as little interference with nature as possible, are current buzzwords. But gardening is always going to involve some tinkering with the natural world. The best we can do won't be perfect, but it might be better. This means adopting organic methods. No pesticides and no fertilisers, but I do still use organic slug pellets and seaweed extract. Purists ban even these, saying that plants become dependent on artificial methods. If the soil is in good heart, it should provide all the nutrition needed and if plants are grown hard, they develop resilience, while increasing diversity in the garden encourages natural predators to take care of pests. So I leave the grass long in the orchard, never dig and try to water less. I choose single rather than double flowers, because these are better for pollinators. It isn't enough, because I love many elements of traditional gardening, but it's better than nothing. If this village is anything to go by, it will take a long time for the average 'keep the garden tidy with a sigh' owner to adopt sustainable practices.

We had a bitter lesson here a couple of years ago, when I was asked to help turn the old bowling green at the recreation ground into a wildlife-friendly garden. The site in full sun was dry and hard, but I reckoned *spinosissima* roses which grow in sand and Mediterranean types could cope with it. Thyme would spread, scillas and poppies would arrive in spring. There were already plenty of wild flowers seeding in the degraded asphalt. It was exciting. Volunteers appeared with cake, and holes were pickaxed for roses. Peter brought a barrow-load of compost from the churchyard. Martyn, the ecologist neighbour, promised to make some insect hotels, and a mixed farm hedge was planted against the road. Aftercare is always the problem with any project, and watering in the hot summer which followed was sporadic. Contractors in charge of mowing the rec decided our patch looked untidy, so they powered over the roses and lavender and cistus. In the autumn we started again, but the majority vote still declared it was nothing like a garden. Too untidy. It should revert to bare ground, which would, the sustainables feared, involve regular doses of poison, even though the roses were growing and the village Wild Flower Group recorded twenty-seven species, including evening primrose and wild St John's wort. But there was to be no reprieve. I should have remembered the lesson I learned from the National Trust. If you plan anything controversial, you need plenty of sessions to prepare people first. Local consultation meetings, more meetings and yet more meetings. We did some leafletting before the work started, but it wasn't enough. The word 'garden' should never have been used. That still has too many connotations of immaculate upkeep.

Mid-March, and the magnolia is out, the grass is growing. Mowing is never my favourite job, but the lawn, which remains a lawn rather than a gravel garden, must be cut. It's light work with a battery strimmer for the edges and a tiny battery mower for the rest. Lawn culture always seems to be a male thing. Why waste time on growing grass in stripes when there are so many better things to encourage? We persuade clients to invest in robot mowers, which roam over their large lawns so often that the clippings don't need to be picked up. Like the robot, I mow in no particular direction, dragging the machine backwards and forwards, which doesn't seem to matter, and the cut grass smells lovely. It will speed up the rotting in the hot bin and I use some under roses as a thin mulch. Too thickly spread and it never rots.

At the allotment, the broad beans have recovered after being squashed by snow, and potatoes and lettuces are in. I halve each potato, even the tiny trio of 'Belle de Fontenay', so there are three rows now. No 'Arran Pilot', bother, so I have had to make do with 'Charlotte'. I like it better than 'Pink Fir Apple', which is too knobbly. Lettuces are planted under the glass barn cloches, which I am frightened of moving, but Alice does it for me. If I pick up a cloche, I always imagine tripping as I lift the unwieldy thing, and that I will definitely fall on shattered glass and bleed to death. Alone. But we love being down here under a high sky where skylarks sing. There is enough white broccoli for both of us, and soon the spinach will be pickable. Alice plants the sweet peas at dusk, cutting our own hazels for props, while I go home to talk to sustainable Jane about a talk I have promised to give in the village in May.

This medieval village is a network of footpaths between and behind the houses on High Street. Before we bought the house, we discovered, at the eleventh hour, that there was a right of way through where the greenhouse now stands. Never mentioned by the seller, it allowed the two neighbouring properties to walk through the garden whenever they chose, with handcarts, to get to Back Lane. Both the handcart rights' owners turned out to be lawyers. One demanded a handsome payment to give up the access. He moved away from the village about a year later. The second legal neighbour, Alastair, asked for nothing and became a firm friend.

The route to Back Lane leads from the oak door at the top of the garden, between Marion's high and Trudie's low walls, to connect with the longer path from Simon and Serena's garden and a door in the corner of Marion's garden. The hard paths are narrow, but there is room on either side for some guerrilla gardening, which I have been allowed by the others to introduce and look after, as long as the way between is left clear. Crocosmia, snowdrops, some ferns, sweet cicely, primroses, the blue bugloss, periwinkle and the odd hollyhock turn the route to Back Lane into an extra slice of green. It is the start of the path to the allotment, or a place to collect heavy loads of compost or gravel in a barrow from a car parked in Back Lane. The wood delivery always arrives via the passage to the top door and gets dumped on the hard standing by the hot bin. The alternative would mean carrying everything through the house. So, it was dismaying to get an email from Simon late last night, to say that Marion's back wall has collapsed on my first stretch of path. When I looked just now it's a huge heap of stones, so my route to Back Lane is currently impassable. Danny and Alan, brothers and masterly builders, will assess it on Monday. They are brilliant and did all the work of making the garden here by hand. Danny is the wood expert. He made the auricula shelter and now grows auriculas at home himself. He cuts hair, sings in the male voice choir and is one of the mummers in the annual post-Christmas play. His brother Alan is a stonemason and roofer, around my age, and keeps threatening to retire. I love an excuse to get them both round for a chat, but the wall looks daunting.

Today, Dan and Huw are entertaining local gardeners to coffee, to meet Midori and Shintaro from Hokkaido. It is, Dan says, a local Hanami moment, a time to celebrate the cherry (*Prunus × yedoensis*), which is now entirely hung with delicate, scented white blossom. On the long table under the open-sided shed looking out over the matchless valley, there are flowers scattered over the plates of cakes and biscuits. In Japan, reverence for each season is traditional. Gardeners are taught 'self-discipline, diligence and devotion'. There is a constant awareness of every season, and the seventy-two gradual changes recorded in the ancient Japanese calendar are prompts to respect the ephemeral beauty of flowers. There is a phrase for each five-day shift: 'The earthworms rise, / The plums turn yellow, white dew on the grass.' I love this poetic observation of the passing of time. In spring, every day is a celebration, with something new to notice.

In Dan's garden we can't stop looking. Ten gardeners, all seeing different plants, all wanting to exchange questions and suggestions, to compare the different giant fennels appearing all over the garden, some feathery, some shiny-leaved, some almost flowering. I like the group of *Gladiolus tristis* under the house, with beaks for buds and grassy stems which flop, but one or two have grown up behind the espaliered pear on the wall and others have been supported by twigs. I'm growing it in a pot in the greenhouse, for its greenish-yellow flowers which are scented at dusk. I think I'll plant it out next month against the south-facing wall under the wintersweet, and I will try to encourage it to grow up through the branches, as Dan has done with the pear. Under a wall joining the path, my favourite scarlet Greek anemones are thriving, which makes me think I could try them on the street side of the house, where the Mexican daisy seeds itself all along the south side of the buildings, between houses and pavement. Ideally from seed, but that will take too long, and Derry says she has plugs coming in next week, so I could chisel them in between the daisies. I ask Huw what the lush-looking spinach is in the vegetable beds. He says it's called 'Viroflay' and that it has been brilliant in the polytunnel below the house. Next year he is going to grow lots. When I get home, I will look it up. 'Monstrueux de Viroflay' is an overwintering spinach for sowing in autumn. It can go under the alarming glass cloches, but I will try leaving some uncovered too. Dan points out a scented daffodil in his new sand garden, a cross between the native *Narcissus pseudonarcissus* and the wild jonquil. He calls it the Campernelle daffodil. That's one to try for its narrow leaves. It's a bit like a smaller version of the scented *N*. 'Trevithian', which I grow at the allotment. It was a great morning for learning more about different plants and for meeting gardening friends.

Last night, Mike and sustainable Jane came to supper. Mike keeps bees and Jane belongs to the wild flower group, draws and sings in the choir. Their son and daughter-in-law have just sailed the Atlantic, but the engine of their boat collapsed on arrival in the West Indies. Mike used to sail his own boat, so he has been advising his son over Zoom. Jane says there is a crochet club in what is now the pottery, where the butcher was when we first came here. Apparently, it meets every Monday at 9.30 a.m. I think you can bring knitting, or any creative hand work, but no pottery so far.

APRIL
Welsh poppy

Bad start to the month this morning, when Jane texts to say that the allotment sheds have been broken into again. We lost a strimmer and a little Mantis tiller just after we moved, but I don't padlock our shed because a determined burglar will get in anyway, so I reckon not to keep anything I mind about down there. Last week, Alice and I took the heavy-duty loppers down to finish cutting the hazels for pea sticks. I left the loppers in the shed over the weekend, but what with fleece and netting all over the floor at the moment, I thought thieves might think such a chaotic gardener wouldn't have much worth taking. I will tidy the shed on a fine day when there is less gardening to do and I have a companion. The companion is needed because it looks like mice or maybe rats have enjoyed using fleece for nests. Anyway, the loppers and a spade have gone, and a nice little potato hoe and my old favourite swoe. They left the oscillating stirrup hoe, which is the best, but I still like the niftiness of the Wilkinson swoe. They also took the canvas director's chair, which is handy to have. Much worse than all these is the loss of the beautiful water bowser on wheels, heavy to push even when empty, so dragging it to a parked car over a hundred yards away must have been a determined act. Bother. I won't replace the bowser, and I will not leave anything in the shed again, but it's a real nuisance taking tools to and fro in the car, when I would rather walk or bike.

The second bit of bad news is that Danny and Alan finally came to look at the collapsed wall, shook their heads and said it would be very expensive to repair. To cheer myself up, I went to collect the *Anemone hortensis* plugs from Derry, where I found Dan buying ferns and a rare Japanese saxifrage. I bought the willow with pink catkins called 'Mount Aso'. I used to have it in the garden and lost it, but it will be better in the allotment for winter picking. I also bought some *Heptaptera triquetra* plants to put in the gravel courtyard. It's a piercing yellow umbellifer. I think umbellifers are almost my favourite plant family, but Peter, a good gardening friend, despises my weakness for what he calls 'another cow parsley'. I love cow parsley and have just one plant here under the damson tree. It's a feature in the May borders at Great Dixter which attracts plenty of complaints about weeds. Plants are whipped out as the flowers go over for fear of seeding everywhere. My single cow parsley is beheaded for fear of spreading, but *Heptaptera* babies (original seed collected in a Bulgarian graveyard) will be welcome. I had a plant before in one of the flower beds, but Derry describes it as a short-lived perennial for dry sun, so I think the gravel will suit it better. Two *Anemone hortensis* plugs went into the gravel, and three more were planted in the gap between the house and the pavement. I had to chip a bit of tarmac away with the hori hori knife to get one in, but the little Mexican daisy loves those conditions, so I am hoping the Greek anemones will settle here.

I added some potting grit to the courtyard, but I think I need at least another inch, which means a bulk bag and waiting until the expensive wall is mended so that it can be barrowed in from Back Lane. This village, with deli, doctor's surgery, gift shop and tea shop, also boasts a fourth-generation family firm with prize-winning lorries and a sideline in delivering gravel. I've chosen a mix of South Cerney stone of mixed sizes, so it looks more natural. If it was a proper gravel garden, Jonny Bruce recommends five centimetres of sand topped with ten centimetres of gravel. The courtyard is currently hoggin topped with a little stone, so I'm hoping I can get away with a less radical approach. So far, the viper's bugloss have survived the winter here, with some *Verbena* 'Lavender Spires', *Verbascum phoeniceum* and one *Dianthus carthusianorum*, and I will add the airy yellow *Verbascum roripifolium* seedlings now in the greenhouse. That's too many plants already in an area five metres square where people have to walk, but I want to see what does best in this new dry and sunny place.

Here, the blossom procession starts with damson and mirabelle. I wish there was room for a cherry, but I chose crab apples instead. Malus have the edge on cherries for having two seasons of interest, spring flowers and autumn apples. The *Malus hupehensis* at the top of the garden has larger flowers and fruit than the multi-stem *hupehensis* from Great Dixter nearer the house. It was one of Christopher Lloyd's favourite trees and this more delicate Dixter version is the true form. Pear and peach follow, and then a hawthorn, *Crataegus persimilis* 'Prunifolia'. Hawthorns, like crab apples, are good for spring and autumn interest. Apple blossom flowers come last, with all six trees turning branches pink and white. There were three trees in the garden when we came, and we added three at the top of the meadow. It was a mistake to buy large trees, and this is something I know and should not have forgotten. Small trees always outgrow maturer specimens, which take at least two years to get going. The 'Discovery', most delicious of all apples when picked from the tree, has hardly grown after four years, and 'Egremont Russet' is far from brilliant; only 'Ashmead's Kernel' is a respectable size and does produce plenty of fruit.

I seem to have missed out on what sounds like a desirable plant. Looking at Sarah Raven's mouth-watering catalogue over lunch, I find a perennial pink cosmos. 'Flamingo', sister to the chocolate-scented *Cosmos atrosanguineus*. Neither are totally hardy but might survive with a mulch and good drainage in warm winters, or can be lifted, like dahlias. Of course it's sold out everywhere. I do look at Graham Rice's New Plants blog on the RHS site, and at new plants in *Which? Gardening*, but 'Flamingo' has, until now, been off my radar. Sarah is clever at spotting new plants worth growing.

Raining again, and I must mow the lawn before I leave for a week in the Mani, on a botanical tour with Jamie Compton. I can't wait to see the garden Tania has made there.

It's hot in Greece, and there are fewer flowers than I hoped to see, because we are here too late. No fields of scarlet and pink anemones, but there are orchids and *Crepis rubra*, the favourite pink starry dandelion, and enormous Pelopponesian cyclamen growing in the walls, with a delicate *Thalictrum orientale*, which I have never seen before. The thrill of the trip is Tania's garden, with an eye-watering selection of plants from Olivier Filippi's nursery near Montpellier. She has been very clever at repeating species in various forms. There are lavenders from dark to pale, and *Teucriums* from misty to glittering turquoise *T.* 'Ourzazate'. But *Teucriums* are bulky birds' nests, unless you prune them as Tania does. Like a clever hairdresser thinning hair, she removes the thicker stalks right down to the base, so that it becomes an airy bush with flowers like blue Adonis, butterflies perched on the outspread branches. I fall for a cistus which turns out to be *C. × skanbergii*, after we spend the week thinking it was 'Grayswood Pink'. Mediterranean gardens are usually short on summer flowers, but silvery Californian salvias take over the later garden, joining silver artemisias and better-coloured forms of the chaste tree, *Vitex agnus-castus*, than the usual washy mauve. These get cut back hard in March. Against the mountain view, there is a billowy hedge of *Medicago arborea*, and everywhere bees are buzzing and butterflies fluttering.

At home after a week away, the wall has been beautifully mended by Jacob, son of Sally who owns the shop, and his dad. Some plants in the passage have vanished but will not be hard to replace, and it's such a relief to be able to get out via the back door again that I go to the garden centre to fetch more grit for the embryonic gravel garden and barrow it in from the car.

The garden seems to have jumped two of the seventy-two Japanese seasons. The Corfu blue honesty was barely registering before I left and the pink magnolia only just starting. Now both are blazing. Primroses and blue *Anemone blanda* are giving way to torrents of cowslips, and forget-me-nots have appeared everywhere, which is fine until they squash later plants, so they will need watching.

A couple of years ago I decided to stop the annual tulip extravaganza. At our last garden, I used to plant thousands in boiled-sweet colours in the big flower beds under the house.

This garden is a very different affair, but I did grow some more restrained mixes in the beds for the first few years. Quite apart from the expense, it seems increasingly unsustainable to use tulips as bedding which gets discarded after flowering. Of course, you can lift the bulbs and dry and store them, if there is room, which there certainly isn't here, but this takes time, and it is tempting to throw them away and just buy more. Before we moved, you could buy a hundred wholesale tulips for £10 to £12; now they cost probably three times that. At the allotment, I have a row of 'Queen of the Night' which come up year after year, and there are other long-lasting tulips, but I won't prove how long they last for me for a few more years. The ones I like, which I want to survive, are 'Mistress Mystic', 'Ballerina' and all the Viridifloras. The Darwins are said to be perennial, but they are a bit too big for my taste. 'Antoinette', which starts yellow and ends pink, has lasted a couple of years in the bed under the apple tree. In grass, the species tulips seem happy, the spidery 'Cornuta', pink *saxatilis*, crimson *hageri* 'Little Beauty', and the one I love best, *T. clusiana*, the striped lady tulip, although 'Peppermint Stick' tends to be offered instead these days. All these have survived in grass, but the one that does best is apricot yellow *T. linifolia* 'Honky Tonk', which turns out to be a better companion for cowslips than flowers of pink, or red and white combined.

Because there is so much in the orchard now, if I want to add bulbs, I grow them first in a pot and then add them to the gaps in spring. This isn't totally satisfactory, because I like a scattered rather than a clumpy look. But the ones that are happy should spread slowly, or that is what I hope. I've made a note to use bulbs to fill a gap in the bed at the back of the greenhouse where there are primroses everywhere. Since the decision to grow plants which make themselves at home, my new policy is to let them be everywhere. Scilla 'Spring Beauty' is increasing, and the Corfu honesty is coming up, so these look like settling. Cowslips are seeding in crowds, in the grass, but I do weed them out of the flower beds. More tiny daffodils would be good. There seem to be fewer *Narcissus* 'Segovia' than I remembered planting and hardly any *N.* 'Kokopelli' this year. Slugs perhaps. Pip, looking in on his way to Wales this afternoon, will probably answer most of my queries.

On the non-appearing bulbs, Pip is reassuring. He thinks they often take a year off, especially if they are newly planted, when they take time to settle, he says. But he looks doubtfully at the dwindling 'Bengal Crimson' rose on the courtyard wall and says that his plant looks just as bad. He suspects they like stronger soil and later sends a picture of a flourishing example outside his mother's house in Wales. On heavy clay. But, but, but: I've just seen the same rose in Greece, on limestone rocks, flowering its head off, and I had it in the last Cotswold garden where it did well after a slow start. I've ordered some Epsom salts to see if that does the trick. If it's happy, this species rose has papery crimson flowers from May until the worst frosts. I have *R.* × *odorata* 'Mutabilis' in the courtyard and that is much more vigorous, and I prune it hard, so that it will regrow fast and flower non-stop. Near the greenhouse there is a second specimen of the Bengal plant that's doing rather well. 'Not the same,' Pip says. About 'Bengal Crimson', those in the know talk of getting the correct form, preferably from Chelsea Physic Garden. The courtyard rose is the real thing, the second one came from a reputable nursery and when I look, I have to admit it is slightly paler.

Gardeners often boast of having a particularly good form of a plant. To the uninitiated the difference is imperceptible. I recently bought a plant of *Coronilla valentina* subsp. *glauca* 'Citrina' to replace one that died. They are short-lived, but I would always want one for its lemon-yellow flowers all winter. Collecting gravel from my favourite nursery at Bathford, I found the same plant on offer, so I bought that too, thinking you can never have enough of a good thing. On comparing them, the local purchase has slightly larger and darker flowers. Both are pretty, but if you look closely, they are not the same.

While we are having tea, Alice gets back from walking the dog home from the allotment. She's off to Venice next week and when Pip hears that, he asks if she will bring back some seed of the wintersweet which she will find hanging over every garden wall. 'It's a particularly good form,' he says, 'I collected some once and gave it to Dixter and they raised twenty plants and sold them all.' Whether Alice will have time to look for Venetian wintersweet is doubtful. My own recent seed-searching experience for *Crocus tommasinianus*, on a bank in a local village, was a failure. In February at Colesbourne, John Grimshaw told me this was the only reliable way of getting the crocus which I love best. 'Collect the seed and broadcast it,' he said, waving at pools of silvery blue. But by mid-April I was too late, so I have ordered a hundred from Shipton Bulbs, who are said to be the only reliable source of true *tommasinianus*. Although, when I get home, on the doorstep is a pot of what looks like hundreds of my favourites with an unbeatable provenance. When, a month ago, Clare from Sissinghurst told me that Troy would be digging some out of the nut walk, and that she would try to save me some, I thought that's never going to happen. But she remembered and I am thrilled. Owning any plant from Sissinghurst is an honour. Vita Sackville-West's gardening articles, which she wrote for the *Observer*, are still worth reading, even if I suspect she was forbidding to meet. Once, when Adam and Sarah were living at Sissinghurst, they let me stay in Vita's cottage. I couldn't sleep one wink, so I prowled the garden in a nightie, overwhelmed by the feeling of a strong and slightly alarming presence.

The 'distressed wood' table and chairs in the courtyard get a frown from Pip. I bought them off the Internet to replace the large oak table, which was falling apart, because I needed something to tuck into a corner, so that the courtyard could become a gravel garden. I knew they were not ideal, but thought they would do until I find something better. Nothing like facing the scrutiny of the best gardener and designer I know, combined with Eva coming to photograph this evening for a book about designers' gardens, to prompt action. Luckily, Alexandra, two miles down the road, sells vintage garden tables and she has a battered-looking green tin table for sale. By eleven, it is in the garden, with a pot of *Anemone hortensis* and a single chair, and it looks perfect. I send Pip a photograph with a caption saying, 'Better?' 'Very much,' he texts back. Phew.

It's a rare client who actually wants to garden. At yesterday's project, reviving a stately shrubbery which was planted in the sixties, the client, who is a painter, follows every step with her iPad, taking notes and sometimes sketching on the screen. When the gardeners, in a hurry to knock off on Friday, leave agapanthus with tentacles coming up for air in the pots, she copies me, plunging the roots down into soaked soil. With bare hands. As we go indoors, where I will stay the night, I say, 'I'll go back again after breakfast and make some more notes.' 'Can I come too?' she asks. So, she does, and we spend two hours looking at what needs doing and what must wait until the autumn. This isn't work but sheer pleasure, to be with someone who really engages with the project and who sees things with the eye of an artist.

Next week it's New York, to see William as Putin, otherwise I wouldn't dare leave the garden again at this time of year. When I was on the NT Gardens Panel, gardeners were never allowed annual leave before the end of May. It is the busiest time for seedlings. I may lose a few, but I hope I will be back for the white banksia rose. I've rather given up on the auriculas, the root aphis is now so vexing that I have lost many of the best beauties, even after drenching the roots in neem and washing the pots with hot soapy water before repotting. Dan was saying he was going to give them up too, they are just too needy. Perhaps the answer is to grow fewer and grow them better, and to use the little lean-to shelves for alpines instead.

Trying to send William flowers for his press night in New York tomorrow proves harder than for an English opening. I like to find a bunch that looks as though it has been picked from a garden, and there are several London florists who can manage to do that. *US Time Out* has a list of Best New York Florists, but all the bunches have titles like 'You blue me off my feet' or 'Orange you surprised', which probably works better with an American accent. Laura rings, and when I tell her, she says, was there one saying, 'I hope you rose to the occasion'? I finally found a florist offering a bunch of lily of the valley costing a king's ransom, which reminds me to go out and see how my own lily of the valley are doing just outside the kitchen door. They spread fast, but they aren't quite out enough to pick, and they wouldn't have lasted on the aeroplane anyway, so BloomNation it has to be. But an email in caps from Kevin reads 'there is a problem with your order'. Too cold for lilies of the valley, he says when I ring, and offers ranunculus, or tulips or freesias. I choose white freesias. Will it be all white on the night?

Last tasks are mowing the tiny lawn, digging out dandelions in the meadow and beheading the only cow parsley plant which is already flowering. I know pollinators like dandelions, but if I allowed any to flower and seed the meadow would be nothing but dandelions. Anyway, bees and butterflies seem quite satisfied with anemones and cowslips.

In New York, Ernie and Arthur offer to drive me to the New York Botanic Garden while William is working all day. I got to know Ernie because he is a snowdrop fanatic. Driving up to the Bronx, spring seems more beautiful than anywhere. Perhaps because William's apartment is fifty-one floors up with views of skyscrapers and a peek of the Hudson River. For days I have seen no green. When we reach the garden, Sam is there to drive us round in a buggy. We get out to walk around various immaculate collections of plants. Lilacs first, spaced widely enough for us to be able to see the habit of a whole bush. I like the looser single forms, especially syringa 'Betsy Ross', with delicate white flowers, a cultivar bred to cope with hot summers. 'President Lincoln' is bluer than 'Firmament', which I have always liked for its old-fashioned, uncomplicated flowers. Lilacs are such summer passengers that I wouldn't plant them in a small garden, but my neighbours on either side here have trees, so I don't need to. If I did want a lilac, it might have to be *Syringa persica*, which is much more graceful and has darker leaves than common lilac. Or, the Rouen lilac, which is another elegant shrub, and both are deliciously overpoweringly scented.

After lilacs, we look at tree peonies. Whoppers in pinks and crimsons with the occasional single white, but some of their glaucous leaves are beautiful. Then we move on to azaleas, banks of them. It's Go Big or Go Home territory over here, and everything is bandbox fresh. Even the underplantings are smart. There are blankets of snowy anemones punctuated by painted ferns. Sam says he wants to show us his favourite tree, and we walk up a grassy slope towards a flowering cherry. Sam parts the low branches and leads us into a cavernous hall of cherry, with a roof of blossom and a floor of tangled branches which spread out from the central trunk. In places where the branches touch the ground, they root on landing, and new trees spring up like columns, supporting the flowery ceiling. On a perfect spring day at peak blossom time, it is unforgettable. The label says 'Fugenzo', a Japanese flat-topped spreading form of cherry. But when I get home, I see that some nurseries here say it is the same as 'Shirofugen'.

MAY
Foxglove

Home is bewitching in the first days of May. In the orchard meadow, cowslips are still flowering, with buttercups coming up fast, and this year there will be many more ox-eye daisies. The wilder area always surprises me. I say 'wilder' rather than 'wild', because I still have the upper hand, but I don't interfere as much as I do in the flower beds. I police the dandelions and goose grass, I introduce bulbs and yellow rattle, Bob mows everything in August and then I wait to see what the next summer will bring. One year it was wild carrot everywhere; another brought too much field scabious. My fault, that version is too big; I chose it instead of the small scabious, so now I am trying to introduce that instead. The meadow area was mown lawn when we came and I assumed over Cotswold brash, so this downland lime lover ought to be happy. It's confusing, because the orange fox-and-cubs grows everywhere, and that likes damp soil. Perhaps the devil's bit scabious, which is also happier in wetter conditions, might be a better choice.

Because this was a medieval burgage town, the plots behind the houses that front the High Street, like mine, all have long, narrow gardens behind them, with a history of orchards and growing produce for the householder. Years of cultivation mean that in places the soil is deep and dark, which is the last thing you want for chalkland wild flowers or Mediterranean types. Cowslips and primroses seed everywhere, and fritillaries increase, but I don't think the frits seed. I must stop trying to pretend this is dry limestone and concentrate on what will be happy. I would like campion to spread, but it doesn't, and jonquils to flourish, which they don't. My dream of scarlet Mediterranean anemones is never going to come true, although they do survive in gritty pockets and the *blanda* anemones are seeding freely in shade, which is what they prefer. The bright magenta *Gladiolus communis* subsp. *byzantinus* is a naturalised garden escape in the West Country where it survives and bulbs increase. Find where a plant seeds itself and you know it is happy. Best of all, last summer, a common spotted orchid appeared, blown in on the wind, possibly from Richard who has a field of them west of me. The prevailing wind is westerly, so with any luck more will drift in. At the edges of the garden, the lime-green *Smyrnium* is getting going. It takes three years to flower from seed and then it takes over. Other gardeners warn against it, but it is easy to remove, and nothing spells spring as well as bright acid green. I've added a bit in the apple tree bed, but it may need watching.

Olivia comes to lunch and crawls round the garden, noticing everything and talking about plants and gardens all the time. Her enthusiasm is catching. After lunch we go to Jo's ex-brownfield site garden, where I try to understand how mowing patches of meadow can delay some flowers while making others repeat themselves. Pastoral management, it's called. There is a thicket of woolly *Verbascum thapsus* growing out of a bed of large stones and one of the best coppiced hazel avenues I've seen. Olivia wants pond lessons and gets down on her knees to inspect beetles in clear water. Jo tells us that goldfinches swarm to the vine pergola every summer, because they like to make their nests in V-shaped branches. She teaches regenerative gardening, so we are getting loads of free tuition. Sometimes I worry that Jo disapproves of my kind of growing, so I tell her I am booked on a regenerative course at Damson Farm. Alison, who runs courses there, often teaches with Jo. They are both members of the eco book club, but it seems I have booked for regenerative veg, which is not one of Jo's courses.

No wisteria still. It's my fault. There was a wisteria on the wall of the courtyard when we bought the house, but it had pale mauve stubby flowers emerging behind brown leaves, so I decided to sack it in favour of something better. At the last house, we were spoilt by an annual display of long racemes of scented flowers cascading down a very old plant of *W. floribunda* f. *multijuga*. Terrific on a three-storey house but overkill on the three-metre courtyard wall. I asked Tania, married to Jamie, who knows everything there is to know about wisterias, and she said, 'Get "Prolific". It always is.' But it isn't and that is because I broke the cardinal rule for wisteria, which is never choose this climber unless you see a flower on it at the time you buy it. They can be slow and cussed; you can wait for five years without a flower and there really is no way to accelerate a reluctant wisteria. 'Prolific' has been fed and watered and trained up the wall as it should be, but, three years after planting, it is barren of flower buds.

Jonny comes to see the garden. He is a walking plant encyclopaedia. He tells me about a euphorbia hybrid called 'Whistleberry

Garnet', which is stronger than both parents. I find *robbiae* dies out quite quickly, so I might try that. I complain about *Brunnera* 'Alexander's Great' being a poor performer. I bought it after admiring it at Lismore in Ireland, among clouds of yellow woad. Jonny says he thinks 'Alexander's' may be overrated and that he prefers *B.* 'Caucasian Carpet', which runs obligingly. My next moan is the disappearance of *Paeonia tenuifolia*. The answer to that is to grow *P.* × *smouthii*, which is a cross between *tenuifolia* and *lactiflora*, but when I look for suppliers it's unobtainable. We talk about growing tree peonies from seed and Jonny says, 'Don't wait until the seeds are black, sow them green. If you wait, they need a cold session before they will germinate.' He leaves me with an envelope of hepatica seeds, with instructions to sow immediately. So I do. I always thought hepaticas were out of my league and only to be grown by John Massey in a shaded greenhouse, but this form is apparently easy. Dixter-trained Jonny, who has worked at the Dutch Hessenhof nursery, is an artist as well as a sensitive gardener. I hope that his planning issue gets sorted soon. His nursery is going to be a valuable way to learn about new plants and techniques.

This time last year Pip and I were in France, where we have been working on a garden in the south. We stay in a castle with round towers, like one in a Babar book. It's Airbnb, but we come so often that Jean-Luc, the musician owner, gives us dinner in front of a fire in the library. The bedrooms are named after French composers. I sleep in 'Lully'. There is no air conditioning in summer, which horrifies those working on the project, but castle walls are thick and Jean-Luc provides fans. All night long nightingales sing outside our windows, and next morning they are still singing. The French contractors shake hands at the start and end of the day, and at lunchtime they wish us *Bon appétit*.

This year it isn't France and nightingales, but a bit of a pause in the gardening life. In *May*. I've been out of breath and suddenly it's worse, so I'm in hospital with an oxygen nose harness and too many tests and no flowers or plants allowed at all. Alice brings fruit and paintings by her friend Louise to put on the windowsill, and I can see sky and the occasional bird above the several-storey car park; Ellie brings delicious food and picks the asparagus and lettuce from the allotment, so it could be much worse.

Home five days later and the *rockii* peonies are over, so I never had the chance to see if the filaments were white. It's news even to nurseryman Jonny that only true *rockii* has white filaments, but since John Grimshaw put it on his Instagram we can all now agonise over whether we have the true form. Then I remember that I had a photo of my peony from last year. Just looked, and the filaments seem to be yellow. Does it matter about 'true form' when it is so beautiful?

The garden has fallen apart a bit in my absence. I love the look of one plant threading through the beds, succeeded by another wave; I prefer that to a collision of sensational incidents. Before I left home, the 'Corfu Blue' honesty and the 'Antoinette' tulips were dominant, but since those flowers faded it looks drab and some of the honesty is overpowering later plants. Laura comes to help look after me and we do some editing of overgrowth. Some of the honesty is left standing for its seed pods, but all the forget-me-nots must come out now, because they are spreading over the important summer plants. The garden looks very different when that is done, with more flowers in waiting than centre stage, but it has to happen. The seedlings of next year's spring will arrive without any help from the gardener and there will still be enough forget-me-nots to dig up to plant in the pots with the tulips by autumn.

May is a gather-your-breath month before the great June moment, which I like to be minor key with gentle colours, more Sarah Price than Sarah Raven. Sarah P's style is so understated it's hardly there. Plants look as though they have grown with no help from the gardener. Sarah P is an artist and everything she does is considered, elegant and restful. Sarah R favours bold, brilliant and thrilling. She confesses she likes her planting schemes to excite. I love both styles, but not at the same time. So Sarah P it is for now, all hesitant half-colours and greenery. In the courtyard the mutabilis rose is getting going with the Benton iris 'Old Madrid' below, and in the gravel I am loving a tall toadflax called 'Peachy', which echoes the limed walls and the rose's weird pinky yellow petals

In the flower beds, umbellifers should provide the background after the honesty. Not cow parsley, except in the purple-leaved form 'Ravenswing', which is less invasive than the wild one, but that seems slow to spread. Derry has an even darker form called 'Dial Park'. There are more cow parsley flowers from *Ligusticum lucidum*, a sort of finely cut lovage from Spain. This is allowed to seed freely and so is *Cenolophium denudatum*, sometimes called Baltic, or Belgravia cow parsley, which likes shadier places. The two classiest umbellifers I grow are *Laserpitium siler*, with tall, strong stems holding white clusters of tiny flowers with plenty of space between each floret, and *Athamanta turbith*, smaller and fussier than the rest, which likes to perch at the edge of the sunny, dry bed under the cherry. Best of all, Derry brings me a plant of *Peucedanum cervaria* 'Purple Flush'. It's going to be enormous, an exotic wild carrot with purple stems and greenery-yallery flowers. But looking it up, it's monocarpic, so I'd better not fall too much in love with it, because that means it will not see a second summer, and at raising it from seed I could fail.

What there ought to be, against the umbellifer backing, is irises, in repeated clumps, but since we divided them last year some have not yet recovered. Annoying. Especially as there is only one group of my favourite pale yellow *Iris flavescens*. The *pallidas* are performing well, although they look suspiciously large, so I am wondering if they are 'true'. I have a few Bentons – 'Old Madrid', 'Quaker Lady' and 'Susan' – and a lovely pale brown unnamed form from a house where I worked long ago. I like the tall miniature bearded types best, with gentle colours and unshowy looks. The prettiest way of growing irises I ever saw was what Julian and Isabel did last year, when they overplanted their irises in clouds of love-in-a-mist. It will be interesting to see if that means the irises get enough sun on the rhizomes to flower as well again. If so, that's a trick many of us might be copying. Instagram is ablaze with lupins and sweet williams, which I love, but not at Sarah Price time, so they are for the allotment and for picking. I've just sown some more of the sweet william 'Electron' strain, which is nearly perennial. The last lot kept going for three years, but they need replacing now.

There is one showy collision which I keep meaning to edit. The rose 'Gertrude Jekyll', in party pink, is next to dinner-plate-sized Itoh peony 'Bartzella' in lemon yellow. In the same frame there is an elegant navy blue delphinium grown from seed. How did this happen? The beds on either side of the path leading up to the greenhouse were always meant to be more traditional than the big bed that connects with the meadow, but this is all much too much for minor-key May. The peony gets picked as soon as it flowers and brought indoors, where it drops its petals crossly all over the kitchen table.

At foxgloves I regularly fail. The ordinary purple sorts seed themselves willingly, but I really want the apricot ones. They were sown last summer at the right time, and lined out in the allotment, but somehow they never grew enough. There are purples and the odd white self-sowns in the garden, but very few tawny pinks. When Catherine and Kirsty come to cheer up the 'im'-patient, we talk pastoral mowing and foxgloves and later I find they have left a present of six handsome plants at the back door, with one in towering apricot flower.

No Chelsea this year because of the breathless blip, so I watch hours of coverage on television, which is mainly tiers of flowered dresses and abnormal quantities of enthusiasm. Derry, in black, looks perfect when interviewed on her elegant and airy stand. I saw her plant collection mocked up in her car park the week before the show, and worried that the hot sun might have accelerated flowering. But Derry is such a cool pro, of course she delivers. Tom's green garden for the National Garden Scheme looks beautiful and real. I think it should have been best in show.

Chelsea is more product than process, more razzamatazz than daily watchfulness, but it's a day I always look forward to, mainly I must admit because it's a chance to meet so many gardening friends and talk to nursery owners about plants.

Many Mays ago, *Vogue* were doing a Best of British shoot to mark the magazine's ninetieth issue, so when the gardeners were photographed by Mario Testino, of course it had to be Chelsea-themed, which turned out to be suitably over the top. Isabel and I ring each other up and wonder what we will wear. I fancy looking like Nancy Lancaster. Isabel thinks tweeds. I call *Vogue* to see if Pip can bring Corby the lurcher, who comes everywhere with us. 'This is Pyrton, Corby. You like Pyrton,' Pip says as we get out of the car. Pyrton is good for rabbits. At other gardens, health-and-safety contractors say 'no dogs' and hand us helmets. *Vogue* is more Pyrton than H&S and says, 'Of course Corby can come.' So, Kim, Pip's partner, asks if he can bring his labrador, but the answer is no. Later, I ask why. Apparently, lurchers are totally *Vogue*. Labs are not.

On the day, we arrive in clean gardening clothes. Early, as models, actresses, cooks and other best-ofs won't get out of bed as early as gardeners. Which, considering we all live out of London, seems unfair. Mario is late. Very, as he has to come by helicopter from Chatsworth. But we don't mind as obviously we are going to be ages in Make-up and Clothes and Hair. Sarah Raven and Isabel Bannerman and I are the only women gardeners, but Jacquetta Wheeler is here too, as the spirit of the flower show, and she spends hours in the make-up van.

When it's our turn, apparently Mario wants us to look 'dirty', so manicurists produce baby oil and clean peat for us to rub on our hands. Then Make-up gets out a pot of rouge, because Mario wants us healthy, and I always look too healthy, so that is a bit dismaying. Sarah and I say, 'Can't we have mascara?' and Make-up says, 'No, you are *au naturel*,' so when we leave the van, Sarah says,

'I'm going to the ladies to put my own make-up on.' But then Mario is in her way, so she asks if we could have eyes at least, and Mario says, 'Of course they can have mascara. How sweet.' We resist naff straw hats, but Isabel is made to wear a pinny, and we are all given brand-new, clean gardening tools. Then we stand in a line outside a black front door, guarded by an actor in police uniform who is frightened of being arrested for masquerading as the Force. So, he disappears from time to time for fear of real police. We are meant to be protesting outside a door which looks like Number 10 Downing Street, and are told to stand to attention and look stern. 'Very serious,' Mario says, 'as though you have been waiting a long time.' Which we have. A Chelsea pensioner, mysteriously in shot, is longing to sit down. Jacquetta Wheeler is dressed in a tweed coat, Marie-Antoinette-type panniers, and a black see-through tulle skirt, with tartan stockings and wellies. The only pair. On her head she has a tiny felt trilby decked with ferns and she carries a huge bunch of flowers. None of us are quite sure what the storyline is, and never do I hear more 'Darlings' in the space of two hours.

The favourite winter flowering cherry has got an invisible worm sickness, or rather something horrible called shot hole disease, which means no leaves on any of the lower twiggy branches. When I look it up on the Bartlett Tree Experts website, I find that 'the most effective management tactic is planting disease tolerant flowering cherry cultivars, such as *Prunus* "Kanzan" [no thanks, too pink by far] and *P.* × *subhirtella* "Autumnalis" [which, in my case, is obviously not disease tolerant]'. Wet winters are accelerators of the disease. I wonder if it's going to be fatal. Christopher Lloyd always said that a death was an opportunity in the garden, so I have already been thinking that if the tree really is failing I could replace it with *Crataegus orientalis*, the silver-leafed hawthorn. Then I can add another cherry for winter to shade the paved area outside the potting shed. I'll wait for Jonny to come and pronounce next week, when he has kindly offered to teach Alice and me how to prune the banksia rose.

You win some and you lose some. In the meadow, the common spotted orchid count is up to five, from none at all a year ago. Aaron gave me one plant last year and another appeared unprompted. This year, there is a second flower emerging near Aaron's present, and the one that arrived on its own has turned into two. Most exciting of all, there is a new plant near the greenhouse. Orchids take three to four years from seed, so these DIY appearances could be from dormant seed, appearing after we stopped mowing the former lawn in 2017. In the war, grass tennis courts left unmown were soon studded with orchids, so I am hoping the first five will be the beginning of many more, although it's standing room only among the cowslips now. My picture of the orchids on Insta gets lots of comments. Wind-blown seed is the general view, and there is agreement that, once they start coming, numbers increase rapidly. Dan writes to say common spotteds have arrived with him after three years and that Donald of Emorsgate Seeds thinks they will be wind-blown, but is surprised to see them so soon. John Grimshaw agrees, but adds, 'I am not convinced that they always take three to four years, but flowering as an annual would be a bit soon.' Orchid seed is so tiny the plants have to rely on a symbiotic fungi for support in the early stages. Looks like that fungus is lurking in the little meadow, so I hope other types might follow. This morning Ellie found another common spotted. Total score: six.

Alice thinks that, in my enforced captivity while I wait to be repaired, I can tame a blackbird. So we lay a trail of mealworms from the top step down to the doormat just inside the kitchen door. And they come, warily at first. Mrs Blackbird is braver than her partner. I tell Alice about Addison saying that he valued his garden more for its blackbirds than the cherries, which give them 'fruit for their songs'. When I re-read his essay, written in 1712, I am surprised to learn how wild Joseph Addison's garden was:

> It is a Confusion of Kitchin and Parterre, Orchard and Flower-Garden, which lie so mixt and interwoven with one another, that if a Foreigner who had seen nothing of our Country should be convey'd into my Garden at his first landing, he would look upon it as a natural Wilderness, and one of the uncultivated Parts of our Country. My Flowers grow up in several Parts of the Garden in the greatest Luxuriancy and Profusion. I am so far from being fond of any particular one, by reason of its Rarity, that if I meet with any one in a Field which pleases me, I give it a place in my Garden. By this means, when a Stranger walks with me, he is surpized to see several large Spots of Ground cover'd with ten thousand different Colours, and has often singled out Flowers that he might have met with under a common Hedge, in a Field, or in a Meadow, as some of the greatest Beauties of the Place.

Here we are, thinking modern gardeners have invented wild gardening. But ten thousand different colours? Really, Addison?

Cherries on the steps and the blackbirds get bolder, especially the paler female. Before I came down this morning, Alice says, there was a marital tussle over mealworms on the doormat. Next step is to sit very still in the chair with mealworms on my knee, to see if they are brave enough to help themselves. The birds have been an increasing delight since we were told about the Merlin app which can identify a bird from its song. We discover linnets, wrens and white wagtails in addition to more regular visitors like swifts, goldfinches, blue tits, robins, sparrows, dunnocks, jackdaws and pigeons.

JUNE
Bath asparagus

M ORE ORCHIDS. I go with Laura and Alice to see a whole field of them around the ex-timber works where Piers and Louise of Garden and Wood now live. They moved their business – they sell vintage tools – from a thatched cottage in Oxfordshire, with a picture-postcard garden, to what looks like an enormous tin shed, all airy space indoors and outside no garden at all. The former lawn, over what was mostly rubble or worse, is covered in thin airy grass and shoulder-to-shoulder orchids. I've seen little bee orchids growing on chalk downland, or in Corfu, or the Mani, but never such tall triple-decker ones and never so many together. There are daisies at the edge of the orchid kingdom and I'm thinking of my own ox-eye cover, so I wonder if they will start to take over here. Piers says they will, unless he stops them seeding. He says that orchids don't like a change in management and as this area has been cut for years, maybe he should go on cutting the daisies and the grass.

At home, the single and species roses are starting. 'Paul's Himalayan' musk is up and over the apple tree and heading for the damson. 'Scarlet Fire' and *hugonis* are flowering and another single rampant climber, whose name I have forgotten, is aiming for the silver-leafed pear. Jonny comes to prune the banksia, which flowers early. It's the double white form, which smells of violets, and it's meant to be less hardy than the yellow banksia, which we had at our last house. The courtyard is my testing ground for tender plants. So far, the rose has survived three winters, because the weather is much milder than it was when I started to garden. Jonny cuts off plenty of branches, creating an airy parasol to float over the courtyard. From Martin's garden, 'Rambling Rector' has arrived, so that rose has to be put in its place too. At our last garden, it tore the roof off the potting shed.

Jonny decrees that the courtyard is too crowded, so he removes the hamamelis in its pot to leave room for the *Melianthus* against the wall. I am reluctant at first, but I think it is right. Sometimes it helps to have another pair of eyes to look and edit. Only problem is, post winter, the *Melianthus* is currently just a woody stump with a minute sprout of leaf. It will recover, but slowly.

Now it's goodbye to home for hospital again, to mend the failing heart. Keyhole is quick, they say, and all the nurses and the doctors smile kindly. I'm lucky. The NHS would have made me wait three months for an 'urgent' review and who knows how long for an 'urgent' procedure. Keyhole it may be, but theatre is theatre, with an uncountable cast of young attendants in green scrubs. It is comforting to find a flower in charge of proceedings. Kerry Ann Dahlia, in what looks like a flak jacket, is at a lectern, asking the lead actor, Jonathan – who by a marvellous coincidence was also Charlie's cardiologist – 'Has she signed the consent?' 'She has,' he says. And I say, 'Yes, I've given him permission to kill me,' because I have. Kerry Ann Dahlia says, 'Right team, you all know what you are doing. Let's go.' And then I hear no more. Later I think, they do this repeatedly and the stage fright must be far worse than for any actor, because each performance involves someone else's life.

It was a week of care and kindness, with daughters and grandchildren every day, and visits from a couple of closest friends, but at last I am home and have got my breath back. Thank you Bernard and Jonathan and Alex in ICU, and Cass upstairs, and all the nurses and doctors at that wonderful hospital with its views of trees and chimneys, for giving me back my life. They have made it possible for me to walk slowly down the blowy garden again, while roses are swarming up trees and daisies are charging across the meadow, where love-in-the-mist has managed to seed itself into a dense overgrowth of buttercups, daisies, grasses, sorrel, scabious and geraniums.

No blackbirds, even though mealworms are waiting for them on the steps outside the kitchen. Perhaps they found other feeding grounds when I deserted them. Or, worst thought, the cats who live on Back Lane have been on the prowl in my absence. It's still a daughter a day, but I'm hoping they can be released once I'm reliably mobile. Today it's Alice, now gone to the allotment to get potatoes, and this afternoon she will plant the campion which I extravagantly ordered before I left. I grew campion from seed originally and planted it, but the pink form has vanished. There is a little ragged robin and a lot of white bladder campion which came from Pippa, but no red campion, which grows everywhere in the hedgerows and verges round here. The jury is a bit out on the magenta *Gladiolus communis* subsp. *byzantinus*, so I thought the bright campions might help by adding more of the gaudy pink which Pippa and Ellie have declared so offensive.

The rain has been good for roses and now is their moment. Although I think single flowers work better in this garden than the showier forms, I can't resist growing a few old favourites. 'Albertine' only flowers once, but is so rewarding in bud and bloom, and 'Munstead Wood' has dark crimson quartered flowers. I'm hanging on to this rose, which has been dropped from David Austin's list. Climate change has made some roses harder to grow, and 'Munstead Wood' is one of the casualties. It took a long time to get going here, but seems stronger this summer, so I will keep it for as long as I can. 'Gertrude Jekyll' is another Austin invention, which is pink and very scented, and 'Guinée', the darkest and I find the trickiest of all roses to please, shares the arch where 'Francis E. Lester' also grows. 'Francis E. Lester' looks like apple blossom, with hips, instead of apples, to follow.

I wish I still grew 'Ispahan' and 'Mme Isaac Pereire' and 'Rosa de Resht', but they are not as healthy as some of the newer David Austin roses. I tried 'Kew Gardens' at the far orchard end when we first came. It's a white long-flowering single rose which makes generous bushes, and I've used it a lot for clients, but when the flowers come out, they touch one another, and I like more space and air around the blooms here. I dislike 'Sally Holmes' for the same reason. I can't resist the fragile alba rose 'Céleste', or airy 'Cécile Brünner' with delicate shell-pink pointed buds. I also have several ramblers, 'Paul Transon', 'François Juranville' and 'Albéric Barbier'. All of these, as well as 'Albertine', are Barbier roses, dating from the early 1900s. The Barbier nursery created new cultivars by crossing glossy-leaved *Rosa lucieae* from Japan with various hybrid teas to produce large scented flowers on climbers with a trailing habit. In this month's *Gardens Illustrated*,

JUNE

Michael Marriott, rose supremo, writes that the Barbiers are still worth growing.

Over twenty roses for a small plot is not perhaps a very modern choice, but I would find it hard to make a rose-free garden. Those with hips make good sense, so *R. moyesii* and 'Scarlet Fire' are both included, and long-flowering roses like 'Cerise Bouquet' and 'Ghislaine de Féligonde', which I am trying as a bush, also earn their keep. 'Ghislaine' was named after the two-year-old daughter of Count Charles de Féligonde, who was wounded in the First World War and is said to have loved growing fruit more than flowers. Most roses, like 'Mme Isaac Pereire', were given the names of society ladies or the wives of nurserymen. 'Ghislaine' is more biddable than most ramblers. Ann in the village grew it from a cutting and gave it to me. It's a repeat-flowering apricot musk rose with hardly any thorns. I can't imagine why I haven't grown it before, and I now remember that Robin Lane Fox used to say it was his favourite rose. I also grow, and have always loved, the repeat climber 'Phyllis Bide', which long ago Esther Merton recommended.

In my forties, when we lived on a hill above the Kennet valley, I went regularly to Esther Merton's open garden days. She was an inspiration. Danish, with a booming voice, she loved food, flowers and encouraging young gardeners. The place was quite wild before the arrival of Sue Dickinson as her gardener, and Esther used to shout, 'Have you come to see my weeds?' to visitors as they arrived. (Sue had the highest horticultural standards and later became head gardener for Lord Rothschild in the Paradise and Plenty Garden.) After Sue started work, the weeds vanished, and there were often days when she taught propagation, so it was a good time for learning. Esther was a generous and forceful mentor. At the time, I was thinking of planting the columnar cherry 'Amanagowa', but she said definitely don't. She also said you can use any flower colour as long as you add white. I've never been sure about that, because white jumps out at the eye, and these days I like things to be more misty and muted.

Just listened to Sarah Raven and Arthur Parkinson, a conjuror on a small scale, on colour for beginners. Neither of them like white as much as saturated Venetian colours, which Sarah sharpens with a scatter of what she calls 'boiled sweets'. Clients run scared of orange and yellow, preferring pinks, mauves and blues, but I love and use sunset shades at home, although I try not to mix them with bluer pinks. The courtyard with limewashed walls is the place for orangey reds. Clients seem to despise salmon pink, but I don't. My favourite of all roses is the evanescent *Rosa* × *odorata* 'Mutabilis', which shifts from pink to yellow and flowers all summer and autumn. The same colours are echoed in the nearly as tall as I am quivering toadflax 'Peachy'. Pam and Sybille, the Sissinghurst gardeners, always disapproved of what they called 'art shades', which is I suspect what I am currently enjoying in these plants. They occasionally came to the last garden and while walking around, they would point and say, 'What are you going to do about that?' It was a question I could not usually answer, because I rarely knew what it was they had noticed.

This morning, I hear the painter Maggi Hambling saying she hates the colour green, as well as alizarin crimson and raw umber. What is it about painters and green? Alice lends me Kandinsky, for his views on colour. 'In the hierarchy of colours,' he says, 'green represents the social middle class: self-satisfied, immovable, narrow.' But I would not want to miss the green of lady's mantle with, for example, crimson campion, *Silene coronaria*. The campion flowers from now until September if it is deadheaded. The double form, 'Gardeners' World', is reputed to be even longer lasting, but it does not self-seed like the single one. Single flowers are better for pollinators, and they never look fussy. Once down the passage, the flower beds are filled with more conventional colours – the blues, purples, crimsons and pinks, which clients seem to like best – although what I call the Sarah Raven phase does not really get going until mid-June.

Now I've started to think about colour, it's annoying, because I don't want to be self-conscious and formulaic at home. But I am interested to discover why I arrive where I do. The answer is that I try things until they jar. There is currently too much pink in front of the apple tree. Pink salvia, pink 'Mavis Simpson' geranium, pink Beth's poppy and about-to-be-pink *Salvia sclarea* var. *turkestanica*. Last year there was acid-green *Euphorbia ceratocarpa* in the mix and some piercing yellow zizia. The euphorbia is no longer, probably due to the wet winter, and the *Zizias* seem to be supine. *Kniphofia thomsonii* ought also to have reappeared. The leaves are there, but no sign of delicate orange pokers yet. I've added a few dark annual cosmos 'Rubenza', still not flowering, and the army of wine-coloured drumstick alliums shows no imminent signs of colouring. Later, there will be blue *Salvia patens* and mauve *Dahlia merckii*, but just now the pink overkill needs correcting. Alice is in favour of a trip to Derry, but that may have to wait a day or two. A scheme of pink, dark red, acid green and blue with Sarah's boiled-sweet orange ought to work, as long as they all manage to be out at the same time.

In any garden where the gardener is engaged, tweaking plant combinations will never stop, but in the days when we moved house frequently, we did more labouring than tweaking. 'When will the garden be finished?' our children ask. They are impatient and clearly bored by how much time we spend outside the new Berkshire house. There are brambles to clear and flower beds to prepare, so every summer evening, once they are in bed, we go back to work, until one of the girls comes out, in pyjamas, to say, 'It's dark, aren't you coming in yet?' But we have seen the barn owl swooping over the wood and are pleased with what we have achieved. 'It's a breakthrough,' Charlie says, after we have cleared and dug and planted, and although it may not yet look like a garden, next year the raspberries will fruit and the roses will flower over arches up the kitchen garden path. 'Very floriferous,' my mother-in-law pronounces, which I recognise is not meant to be a compliment.

Change is what makes gardening interesting, but it's often hard to tell clients that weather and rain can undo what you plan and that you can never rely on plants to deliver on time. On the NT Gardens Panel, I remember learning that even at Sissinghurst eight per cent of the plants were lost annually. Shrubs, or modern planting schemes which rely more on grasses, are more reliable, but fleeting is what I like best at home. In a small garden every season matters. Quite often clients are away, or not interested in a garden where you have to search for flowers in cold weather. Everything coming up roses is what most clients want. June is *the* garden month, the time when everyone expects their garden to be best. Eva Nemeth confirms that nearly all of us want our place photographed now. Today she spent with Dan in the valley, followed by the evening with me, and tomorrow she will be with Sarah Price.

It's open house for blackbirds just inside the kitchen door again. Sparrows have also learned that there is a running buffet of mealworms on the doormat. I worry that if birds come too far inside, a sudden fright will make them forget where the exit is. But doormat feeding has to stop now, because one rat, which usually means many more, has discovered there is free food on offer. It's apparently an urban myth that you are always only six feet from the nearest rat, but this is a rat too close for me. William arriving from New York tomorrow may suggest poison. I would rather not, as it's lethal to birds too, so the taming of blackbirds is out from now on. It was a lovely distraction in the first idle days out of hospital. We learned that blackbirds prefer cherries and grapes to strawberries, and that sparrows like eating in company. Female blackbirds are braver than males, and will take a grape balanced on a shoe or a knee, and once from Ellie's hand. And in the spirit of new rules needing to be broken, we put out mealworms on a supervised plate at breakfast. Alice and I spend a good ten minutes watching a mother blackbird feed her teenage offspring with repeated helpings of mealy muesli. She crams her beak with food while the fledgling stands by, open-beaked and cheeping, then eats everything on offer. Sometimes the male appears, but never helps or feeds. Guarding, perhaps?

Pip sends a picture of the airiest *Verbascum roripifolium*. Flowering. Which mine aren't. I gave him a plant I had sown from Derry's seed, after I admired it last summer. I kept the seedlings in the greenhouse all winter and planted mine out in the sunny courtyard gravel, where they flatly sit, showing no sign of growing into the graceful pinnacle adorned with lemon-yellow butterflies which Pip is enjoying. His plant seems to be growing out of a mass of ox-eye daisies, but Pip can make anything flower. Later, Derry admits she has raised fifteen spring-sown seedlings, but so far only one has flowered, although she thinks they might all make it this year.

Dan and Huw have a birthday party in a tent on the hill beyond the garden. As the sun sets below the landmark trees at the head of the valley, a hot-air balloon floats into view above the horizon and sails away over the side of the hill. I don't know why it is so moving, but it is.

Ellie drives me to the allotment, where I have not been for over a month. I pick broad beans and sweet peas, both better crops than I've ever known, so all that spring rain must have done some good. We cut branches off the blackcurrant bushes, and I sit in a chair to strip them. This simultaneous prune and pick is by far the best way of dealing with blackcurrants, as long as the berries are ripe. We cut some branches which are not quite ready, which is a bit of a waste. Ellie picks more strawberries, digs some potatoes and plants some overdue leeks which are crowded into tiny cells. Dowding says seedlings should go out at three weeks, but these have been left for at least double that. I pick the outer leaves of the Morton's secret mix salad from Real Seeds. It's the best lettuce to grow for cut and come again if, like me, you are slow to make successional sowings. Although it's hard not to grow a few 'Little Gem' and 'Black-Seeded Simpson'. We forget to pick the globe artichokes – there are plenty of those again. When Derry looks in, we give her some broad beans because there are so many. Later, she texts to ask if the reddish ones have a name, as they taste so much better than the ordinary ones. It hadn't occurred to me to try individual tasting. I think I ordered the seed of 'Karmazyn', because I thought the flowers would be crimson, which they aren't. It is the beans which are coloured.

JULY
Crested moss rose

CATHERINE AND KIRSTY and I are booked to visit Tom Stuart-Smith's Plant Library. It means leaving at eight, with Catherine at the wheel and the wheelchair in the boot, because Hertfordshire is two hours plus away and we want to be there soon after ten, when it opens. The Plant Library shows dry zone plants in sand at the top of the slope, wet zone ones in the lowest area planted in earth and a few in-betweeners in soil in the middle ground. Our first thought is that the dry plants, all airy shimmer, are much prettier than the wet types. The only wet plant I want to grow is the sparkling white *Thalictrum* 'Splendide'. There are masses of plants I do not know, but you can scan a code on a post at the corner of each block. Call me old-fashioned, but I'd rather have numbered beds with lists online. It would be good to be able to refer to everything later. Tom appears and tells us that he wishes he had made the sand deeper: 150 mm is apparently not quite enough. Kirsty says that too shallow is risky. She knows because she tried in Devon.

I don't take nearly enough notes, and the sandy paths mean the wheelchair keeps getting stuck, but I enjoy sitting in front of the beautiful barn designed by Tom's architect son Ben. There is

Salvia arizonica, a ground cover in shade, and I am surprised to see a good group of *Gladiolus* 'Ruby', also in semi-shade. Here it flops, I imagine because I have made the soil too rich, and it has been known to pack up in a cold winter.

After Kirsty and Catherine have looked at every single plant, we head to the Barn Garden, where the Mount Etna broom trees are cascades of yellow above the restrained blues, purples and silvers planted in the sunken garden below. Later, I realise that I don't think Tom ever uses red in his schemes. We are invited indoors to eat our picnic with Tom and Sue, which is definitely the icing on the cake. After lunch, it's his sister Kate's garden, a wildly romantic confection of hazy plants crowding the narrow paths. The verbascums aren't quite out. The pale spires piercing the misty colours must be an incredible sight, but it's a thrilling experience even without them.

Catherine drives us back to her house and Laura and Alice come to collect me. They can't resist swimming in the pool of pools, surrounded by plants, with a view of the valley beyond. Catherine's garden has tumbling roses and a hedge bedstrawed meadow, and a much better version of the Etruscan honeysuckle than the one I grow called 'Michael Rosse', which is more compact with yellower flowers. Another case of 'better form'. Vita Sackville-West once wrote that if she had a small garden, she would want to make sure she grew only the best form of everything. But what if you only discover the 'best form' after you have already planted an inferior one?

Working gardeners Aaron and Matt come to lunch to talk about making a garden in Corfu, where I have been working for years. Except it is not so much a garden as an intensification of a beautiful place. They bring the lunch, flowers and a rare wild rose from China, *R. indica*, grown from a cutting, a relation of the banksia, with clusters of white flowers. I must find it a host for it to climb at the edge of the garden. They tell me to take out the enormous Etruscan honeysuckle now scrambling over the 'Francis E. Lester' on the arch at the far end of the garden and plant 'Michael Rosse' instead. But I probably won't. Rose control will have to be outsourced to Jonny, or Alice, or Jane, as I don't do ladders any more.

Aaron and Matt also bring a bunch of single dahlias, grown from the seed of *Dahlia coccinea* which bears my name, cross-pollinated with other varieties. They are clear-coloured, large, open-faced flowers, and, unlike mine, their petals do not flop. Matt says I can choose any I like. I must ask if they are likely to flower as early as my *coccinea*, the only dahlia out for the last fortnight, just joined by shocking-pink 'Winston Churchill'. But no favourite 'Waltzing Matilda' yet, nor 'Karma Fuchsiana', 'Verrone's Obsidian', nor any of the starry Honkas, because slugs are still eating any new growth. Officially, I only grow single dahlias, because they are better for pollinators, although I can't resist a few with petals as pleated as a Fortuny dress. This rainy year has been impossible and there was a moment when I thought the dahlias would never make it. Richard writes that his, in pots, protected by copper rings, have all been destroyed. I have noticed that the varieties with dark leaves, like 'Waltzing Matilda' and 'Magenta Star', seem less troubled than those with green leaves. The other pest of the moment is black fly, clustering on stems of the species *Dahlia merckii* in pots outside the kitchen, but they are easy to rub off with finger and thumb.

Ellie is definite that this is the last moment for sowing winter salads. So we do. The radicchios 'Rosso di Verona' and 'Variegata di Castelfranco' are old favourites, but I'm also trying some mustard called 'Dragon's Tongue', which can be cooked when the leaves get too hot to eat raw. Parsley sown twice this year and planted outside has been eaten by slugs, so I'm reduced to growing it in a pot in the greenhouse. Basil, also in the greenhouse, has been discovered by slugs, or snails. Ellie says they hide under the pots, but we can't see any. The other slug fest is chrysanths, at any stage of their growth, although the ones at the allotment seem somehow to escape.

In May, I Chelsea-chopped half the stems of the big 'Prichard's' campanula. Now the uncut stems have huge trusses of flowers and next to them, at the same time, the cut stems are flowering sparsely. I thought the point of the Chelsea chop was to delay flowering by reducing alternate stems of perennials to a third of their height. What I didn't chop was the tiny pink-flowered mallow, *Althaea cannabina*, now eight foot and still growing in all this rain. Jo writes, 'It doesn't look very good when you chop it halfway down. I've taken to cutting it to the base and then letting it grow again. Or cut every other stalk out.' This summer of rain has meant that plants like phlox, which usually struggle here, have grown really well. So too have the *Dieramas*, the angel's fishing rods, which like moist summers and mild, dry winters. I only have the pale and a darker pink form, but I would like to grow 'Blackbird', which is dark as a damson rather than the colour of a male blackbird. Derry says they only survive for three years with her, but I think that in some places the soil here may drain better than hers does.

JULY

Hollyhocks are enormous this year and all the lower leaves are rusty. I pull them off, leaving the plants with stems like giraffes' legs. But I now wish I had not left the one on the corner of the path, which is taller than the apple tree and has nothing to hide its four-foot stem of naked green. Self-sown seedlings usually appear in the path and I never know what colour they will be. Some I leave and some I move further back into the bed, which given the naked stem problem is a better place for them to be. All mine are single. Pip gave me seed of his very dark form *Alcea* 'Nigra', but now it is paler, having crossed with pinks and purples. Then from Pippa I had a clear pink collected on the Île de Ré, where rust-free hollyhocks line the walls in a beautiful range of colours. But colours cross, which means some murky purples and a few disastrous pinks. I've just pulled out a dull magenta in the big bed which looked horrible with the scarlet of clear red 'Hellfire' crocosmia and the crimson *Dahlia coccinea*. The only single-colour seed available seems to be 'Nigra' and a pink, or the halo strain with pale centres which I like less, so if I want to be sure of getting the colours I like, I need to try taking root cuttings next spring, which means labelling plants before the flowers fade. And if I want rust free, Sue recommends the 'Happy Lights' strain. But these are not as bright as the classic hollyhock, because they are crossed with the form with palmate leaves, which mostly escapes rust. This is *Alcea ficifolia*, a clear lemon yellow. I have that and like its slender habit better than the sturdy stems of the cottage garden favourite. *Alcea rugosa* is another delicate greeny-yellow rust-resistant hollyhock for a dry spot. Jonny the Knowledge says that it is critical to keep all hollyhock seedlings watered in the early stages. On the Île de Ré, where we once holidayed, they grow in every chink between the houses and the pavement. How much water do they get there, I wonder. Gardening is a puzzle.

On Instagram Tom writes, 'Most of the time I think of ecological character, scale and feeling first – then texture and form and colour follows well down the list of priorities.' Which is all fine, except when a client wants plants which don't suit the soil, and is less keen on form and texture than colour. At home, I put feeling first, and I try to respect ecological concerns, but I do love colour, and I would always want a garden for all seasons. I think about this advice because today Kirsty and I must work on borders for a client who likes a grand and showy effect. 'So it can't just be gauzy,' Kirsty says. And I wonder if grand and showy is best left to others. Gauzy and wild is what I like best. Julian and Isabel of course would deliver the perfect scheme. But with current low energy, I am a bit defeated by imagining this revamp of grand borders on unfamiliar soil, which is why Kirsty is roped in to help. She is an artist with a subtle eye. Her plant knowledge is formidable, and her style is original, so we sit at the kitchen table toing and froing about our choices. She is insistent that we should include a banana. 'For scale,' she says. But I think this client likes trad better than exotic, so no bananas and no grasses is what we must do.

One of the problems with making borders for clients is of course the maintenance. Will there be a skilled daily watcher who can tweak and fuss and deadhead the plants as we both do in our own gardens? Mostly the answer is no, there won't. I ring the head gardener to ask if daily watchfulness will be available on this particular job, and he sounds doubtful. In the past, gardeners have only spent a day a week in an area of double borders eighty metres long. I doubt this will be anything like enough maintenance for my proposal.

At the end of the long day Jonny looks in and astonishes us with another horticultural fact. Salvia 'Amistad' and indeed all the *guaranitica* tribe need much more water than most other salvias, he says.

Looking at the winter-flowering cherry, which I had decided to condemn when I saw that the lower leaves were riddled with holes, I notice that there are still leaves at the top of the tree, but many brittle twigs lower down. I think I will keep it another year; it could recover. In case the shot hole proved fatal, I did order the very hard to find oriental thorn. It will be a small bare-root whip, so I could plant it to shade the area outside the potting shed. If the prunus does die, I can always move the hawthorn to fill the gap and plant another cherry near the potting shed. Old wisdom maintains it is better to plant something different after disease strikes.

Dan and Huw have a summer tea party for local gardeners. The table in the loggia is laid with cakes, scones and a proper wobbly jelly. The new sand garden is filling up and the wall that backs it finishes the scheme perfectly. There are nasturtiums climbing the asparagus and, in the huge borders to the south of the house, *Gladiolus* 'Ruby' the colour of Huw's jelly has seeded everywhere.

Luckily for me, July is an easy month for the gardener. I thought it would take six weeks to recover, but apparently twelve is more likely. Deadheading and watering the pots and greenhouse every evening are the only critical tasks and, if I water with a hose rather than with cans, it is hardly strenuous.

The tomatoes are very behind because they spent too long in small pots while I was away, but now they are beginning to climb the string which is attached to the highest point of the greenhouse. They look much better this way, more flourishing. Later they will make tented arches like the ones in the Eric Ravilious painting, which is the inspiration for how they are grown at the Paradise and Plenty garden and now copied by me. The string needs to be the thick, strong kind, like miniature rope, and as the tomatoes grow I twist them around their support, taking off the lower leaves and any that shade the fruit.

When I did the Dowding 'no dig' course, Charles told us that the only seed of tomato 'Gardener's Delight' worth growing is offered by Real Seeds as 'Gardener's Delight – small but super-sweet Irish version'. I think we had all begun to notice that recently the flavour of this favourite has been disappointing. Ellie likes 'Yellow Cherry', so I am growing that too, as well as an unnamed Waitrose plant of a black tomato, currently ahead of the others. I've potted on the rest of the seed-grown veg from the smallest cells, although the Ellie/Dowding diktat decrees they should go out as tiny three-week-old plants. But the allotment is over the main road and tiny plants tend to suffer when I only get there a couple of times a week. The snag about growing tomatoes in the greenhouse is that I have to turf them out once the half-hardies need to be taken indoors. The season would be longer if the fruit had a better chance to ripen. I am tempted, extravagantly, to get a polytunnel for the allotment.

The newest addition to the shop of shops is bags of mixed salad leaves, which include marigold and nasturtium petals. My second batch of Morton lettuce mix is still going strong and there is a third row which should see me through September when these are done, as long as I only pick the outer leaves. It's good to know that I can get a bag of home-grown salad if I can't get to the allotment, and when I ask where the shop ones are grown, it turns out that they come from a field over the road, where Charlie, whose parents live in the village, has started raising organic veg to be sold here. I write a tiny column for the village magazine, and when I contact Charlie, he suggests we meet on Friday afternoon, after he's packed and delivered his weekly pickings to the shop. He tells me he took a career change from hospitality after lockdown, and went to work for Charles Dowding, the grower who has revolutionised veg gardens. Last year, Charlie took over some land the other side of the main road. He lives over half an hour away, so it's hard to imagine how he finds time to manage the new enterprise in addition to his other work. But he does and the patch is immaculate and enviable. He's hoping to make a go of it and the salad bags are a start. It looks as if there are heritage tomatoes to follow, as well as beetroot and onions.

When he took over the plot, he cleared the ground by putting down a layer of sheep's wool and two inches of compost, and then he grew a crop of potatoes through black plastic. The Dowding technique involves adding a couple of inches of compost a year, so Charlie buys in green waste and bark to add to the stable manure from his mother's horses. He saves a lot of his own seed and when slugs were really bad earlier this year, he would pick them off the crops after dark. If you total the labour cost, the time to grow, pick, bag and wash the veg, as well as the price of compost and seeds, his salad bags are a bargain. Lockdown was dire, but it did make people stop and think about their lives, and some enterprises, like Charlie's, turn what was once dire into what is now inspiring.

Laura and I go to Franklin Farm, where Kim and Pip have created a bucolic idyll. You bump up a long track with wild flowers on either side, park the car behind a small barn and lift the latch of a wooden door, to find a secret place overflowing with Longhorns, turkeys, hens, hollyhocks, lilies, lemons and flowers everywhere. There is a delicate small meadow, with scabious which has not fallen sideways among the thinnest of grasses, because meadows are better on chalk than any other soil. There are knapped flint paths laid by Kim, and the greenhouse is bursting with rare cyclamen seedlings. Lunch in the cart shed is home-grown salads and strawberries in pale pink jelly. Pip is as good at cooking as he is at gardening, farming and playing the piano, at the same time as being a full-time successful designer. 'Goals,' the younger generation might say, which is probably a better reaction than outright envy. The last July outing and the best.

AUGUST
Poppy seed

SALLY WHO HELPS to organise working parties in the churchyard wants to meet me there. I used to be involved in various working parties, but it's more for advice these days that I am needed. The look of the ground around the church is the result of community labour. Nobody gets buried there any more, but there are some old tombstones near the building and a lot of grass, much of it on south-facing slopes with a view of the valley. All of this used to be closely cut, but Peter the compost maker was instrumental in getting a more sustainable approach adopted, and now the grass banks are left to grow wild flowers which are scythed once a year instead of mown. There is already a well-maintained approach to the main door, via a pair of flowery borders, but a bed under the wall has been neglected and in another corner the Garden of Remembrance is overgrown and in need of attention.

Sally has transformed these areas with help from volunteers who come and go, but she wants me to look at how we might improve the experience of being in the Remembrance Garden – a mown space with seats between two semicircular beds. The problem is that both beds have the same inner curve so that their edges seem to line a wide path that you pass through, rather than creating a circle where you want to linger. Sally is a bit dismayed at the idea of changing the edge of one bed from convex to concave, because it means digging out hard-won planted ground and replacing it with grass. We are going to look at it again in the autumn, using a hose to set out the new curve. If the space in the centre of the beds was rounder and the entrance points narrower, it would make the garden a more private place for people to sit undisturbed. The other Sally, Sally the vicar, had suggested that it might be nice to include rosemary somewhere, but the garden is shady, so we wonder if adding rosemary bushes on either side of the sunny entrance might enclose the space and narrow the entrance. Although privately I think evergreens might be better in this particular setting. More thinking will help. It's always good to have time to make decisions about changes, which is not generally possible when making gardens for clients.

Before leaving the churchyard, I take a look at the cutting patch we made five years ago to provide flowers to pick for the flower arrangers. 'Grown not flown' and no oasis at all has been preached by Shane Connolly for years, so this is another sustainable aim. Creating the plot was a major task at the start, because brambles had to be cut down and then hacked out with a mattock. Digging out their roots is harder than levering them with a mattock, which looks like a pickaxe with a broader blade. Because brambles put down roots with every stem which touches the ground, that's a lot of mattocking, so when clearance had to be done, four of us took it in turns to swing the heavy weapon. It's a satisfying moment when the root is scooped. After bramble removal, the ground had to be tamed by cardboard and compost, which meant collecting old boxes from the shop to spread on the soil, topped with a five-inch layer of compost. When the cardboard rotted, which it did after about six weeks, we pulled the compost to one side and put down some more flattened boxes. The plot was useable after three months, although there were still quite a lot of bindweed roots to be wrangled out of the ground. The best thing about the 'no dig' method is that the weed roots work their way to the top in search of the good stuff, making them easier to pull.

The original plants I supplied were mostly deep red dahlias and blue flowers, but the arrangers say that pale colours stand out better in the dark interior of the church, so some repeat-flowering roses and alstroemeria were added later on. There are some new dark Peruvian lilies which I tolerate, like 'Indian Summer', but the paler sorts always seem a bit too floristy for my taste. The pickers put a stern sign on the bed that reads, STEMS MUST BE PULLED, NOT CUT, which is wise if you want to keep picking alstroemerias. But if you want one cut flower to pick all summer, nothing beats cosmos. We used to grow plenty of the white 'Purity', until the work involved in annuals became too much. I have less time to devote to keeping the plot in order now, so the flower arrangers keep it tidy, adding plants from their own gardens which they enjoy picking.

The zinnias in the allotment have finally got going. I always sow them late, around the second half of April, or even early May, because they dislike having their growth checked and they can't be planted out until the weather warms up. It hasn't been a summer for zinnias, but they are flowering now in dusty pink, and orange and limey green. I like the look of their matt petals, so different from the varnished flowers of dahlias. Dahlias also like sun, but, unlike zinnias, they need plenty of water. There has been no shortage of that in this rainiest of summers, but wet is what slugs like too, so the dahlias have been decimated. I miss them; they are usually the late summer mainstay in both garden and allotment. What to do about slugs is apparently the most asked question in any garden forum. Birds, frogs and hedgehogs will eat some slugs. Grit around plants can help and some swear by coffee grounds, eggshells or wool, because even certified organic slug pellets may damage wildlife. I do still use a few, perhaps five specks per plant, for vulnerable seedlings and fleshy stems like dahlias. Ideally, plants would be grown hard and only put out once the tender early leaves have toughened, but there is never enough room in the greenhouse to keep plants beyond infancy. Nematodes are a bit complicated to use, because the soil needs to be warmer than 5 degrees Celsius day and night, and they only last for about six weeks. You need to get the timing right before shoots appear, but if you do use them the slugs are miraculously devoured by minuscule worms, so there are no corpses to poison the creatures you care about.

I've ordered my bulbs. I always hope to get it done in the first fortnight of August. Narcissi need planting in September, and to get bulbs in time you need to be quick, because deliveries get sent out according to when the order is received. This year I should be miles ahead, because in April I ordered jonquils and the impossible to find *Crocus tommasinianus* from one of the few nurseries who do sell the true *tommies*, the little silvery self-seeding flowers which I love and have missed since we moved here. I also order a few tulips, which I hope will be perennial, although last year I meant 'Antoinette' to be a keeper but then couldn't bear the sight of brown stems dying in the flower bed. So I dug the bulbs and dried them, and now I must take them out of the string bag where they have hung in the potting shed all summer and plant them in a pot to see if they did survive. In case they fail, I've ordered more to put in the beds, as well as a hundred favourite 'Ballerina'. There is a problem with this lovely orange-scented tulip. There seems to be a false version around which flowers a bit earlier and is distinctly less lovely. I'm not sure which one I will be getting this year, but I'm hoping it will be the 'true' form.

Bob next door, best of neighbours, has mown the meadow for me with his heavy-duty petrol strimmer. My small battery strimmer chokes on long grass after a few minutes and, by mid-August in this soaking summer, the grass is as high as I have ever seen it. In past years Bob has returned with his mother's mower to finish the job, but this year the strimming reduced the meadow down to patches of scalped earth. Wild flower meadows need drastic haircuts if they are to produce more flowers than grass. When I ask how the scalping was achieved, Bob says he decided to top the grass first and then to take it down in several swipes. For once it is not raining, so the cut grass can dry for a few days and then Laura and I rake it and carry most of the hay to the big compost heap at the end of the garden. The removal of grass is vital; left on the meadow it will enrich the ground as it rots, which will favour grass over flowers. A week later the ground gets a second, fiercer, wire-tined raking from Alice. This is to remove any grass left and to discourage moss and thatch. The bare patches are now even barer, which means they are in perfect condition to receive the yellow rattle seed which I collected in June.

We know now that yellow rattle is the wild flower meadow maker's ally, because it weakens the growth of grass by feeding off its roots. Rattle is an annual, so some will sow itself each year. In case it fails, I like to scatter some from collected seed once the grass has been cut. It needs to be very fresh, so a spring sowing from bought seed is less likely to succeed. As well as rattle, I've had an orchid stem upside down in a mug on the dresser for the last month, waiting for the seed to drop. It is now lying at the bottom of the mug and is about as visible as white pepper. I am going to sow that in the bare patches, mixed with some sand so it doesn't fly away.

The garden seems much smaller once the meadow is mown, and this is the time of year when I am reminded of the apple tree mistake I made when we moved. The three trees, 'Discovery', 'Egremont Russet' and 'Ashmead's Kernel', are my favourite eaters, but only 'Ashmead's' has flourished. Raised in Gloucestershire in 1700, it seems to be happy in the county of its origin. 'Discovery' has sulked and hardly grown, which is not fair as one of its parents is 'Beauty of Bath', and Bath is my nearest town. Apples are always said to do best on home ground, which is why many of them have names like 'Worcester Pearmain', 'Devonshire Quarrenden' or 'Flower of Kent'. 'Egremont Russet' is far from local, because it was raised in Cumbria and here looks distinctly unhappy, with branches more like a topknot than a spreading canopy. Even if the location was suitable, it's my fault if the trees are forlorn. When we moved, I decided to buy big trees, standards, which we could walk beneath before too long. They arrived at over four metres tall, not in the best of shape. It's easier to transport big trees if their crowns are small, so they tend to be sent out with lollipop heads, otherwise the branches can get damaged in transit. Added to this, any large tree stands still for a couple of years before it starts to grow, so buying smaller sizes means growth is faster and trees end up stronger. Big trees need a lot of support, so our trees needed stakes. Huge ones need underground guying, but young trees under two metres tall are good at adapting to wind by themselves, because it stimulates them to put down more roots.

I know all this and often try to persuade clients to opt for smaller sizes of trees, so why did I not follow my own advice? The Ashmead's Kernel apples are good, but the Egremonts are tiny. My fault again, I should have thinned them. Sometimes this happens naturally in what is called the June drop when little fruits fall to the ground. But I was away in June, so who knows what happened this year. There are a few apples on the tree, which is still lacking a fifth main branch, so it looks lopsided as well as stunted. Apples need open centres, so we prune out anything that grows inwards, aiming for a vase shape. Old pruning advice is that you should be able to throw your hat through the middle of the tree for light and air to reach all the branches.

Every year around now, I wonder if I should axe the forlorn russet and start again. Its trunk is very rocky, which means it can't have much of a hold on life. Next year I'm going to remove the juvenile apples from the sulking trees, so that they concentrate on growing. This is their last chance. Orchard trees are a surviving feature of the burgage plot gardens in this medieval village. We inherited a couple of cookers when we moved and, on either side, the neighbours have damsons and more apples. When people lived and worked at home a productive plot was what they had, rather than a sterile lawn with tidy flower beds. I worry that gardens for most people are a chore, just outdoor housekeeping rather than nourishment for both body and soul. I wish it wasn't so.

The ripening of the first apples always used to be the signal for a family holiday on the northernmost Cornish coast. We took baskets of fruit and vegetables in a car crammed with bodyboards and terriers. 'Are we nearly there?' the children ask, for most of the long journey down. I still love the sea and this year Ellie and I have three days in a familiar cottage. The water is warm and at low tide we walk as far as a rock where a wreck sticks out of the sand. Samphire and thrift grow on the cliffs, fuchsias, montbretia and blackberries are all along the lanes and in every front garden there are enormous mophead hydrangeas. Local distinctiveness is something I used to think about a lot. Gardens are samier now. We grow what everyone else is growing rather than what does best in the neighbourhood. I have planted hardy fuchsias and various crocosmia at home, reckoning this is almost the West Country where they belong, but the 'Limelight' hydrangea, which was here when we came, is the only one of its kind. It has now been moved three times, and I am still wondering whether it should stay, because hydrangeas are hopeless in dry years, and to me they never look quite right planted in the windy uplands. 'Limelight' has been fine this soaking summer, but the Cornish hydrangeas always look much richer than any round here, I suspect because of the perpetual mild damp. I could get the same results by copious watering, but once plants are established, I increasingly think it isn't sustainable to water ornamentals regularly any more. Pots have to be watered but grouped together they stay damp for longer. I still have too many small pots clustering around bigger ones, and in hot weather the little ones might need watering twice a day.

Vegetables must also be watered and that is a heavy chore because there is a 'no hose' rule at the allotment. The water trough is very near my plot, but even then, it is testing to carry two full cans thirty yards to the top, where the celeriac is planted. No rain for a fortnight means most crops will flag. Some, like celeriac, always need a drink. Potatoes and beans need plenty after they flower, to make them swell. Salads and brassicas grow better for lots of water, and everything needs a can after planting until it's established. If there is wood chip available, I use that for mulching the thirstiest crops, but I dread a drought, because each row takes at least three cans and four would be better. I thought the beautiful bowser, the water tank on wheels which was stolen, would make life easier. But watching me trying to push iron wheels over grass and quite often spilling half the load, the daughters used to be scornful. 'Just as well it was stolen,' they say.

In other years I have gone abroad in August, but travelling still seems too daunting, and staying up late is out of the question. I miss Corfu where I spent so many weeks planting, making a garden that was not a garden, because the place was already too beautiful to change. 'No bougainvillea?' visitors asked when they came to drinks, seeing only native plants around the house. There were gardenias and plumbago in pots in the courtyards and a kitchen garden with artichokes and oranges and some zinnias for picking, but very little else, except in spring when there were wild flowers everywhere. In summer I was a guest, although that often involved some pruning of branches in the view, or discussing the moving of beautiful objects for another visionary project, but there was always plenty of time for swimming with Catherine my artist friend, before she spent whole days painting in the olive grove.

Long, long ago the Venetians planted olives at Kanonas. In a ceaseless game of grandmother's footsteps, time has crept up on the trees, and the wind has fixed their gnarled trunks into twisted poses. Some are frozen in dance, while others face the sky, or turn to look down at the bay. Their shimmer of leaves frames patches of azure sea above and cobalt sea below, and all day long the forever sounds – the whispering, sighing, clicking, cracking, buzzing of unseen lives – form an undertow to the daily passage of people. The pruners, the searchers for spring flowers, the walkers at dawn (or dusk in the cool of the day), the summer swimmers, the olive gatherers; all pass beneath the branches. And as they do, they slow to another timescale.

Just risen over Albania, the sun makes a silver pathway almost too dazzling to travel with the eye. Down in the little bay, Catherine and I lie on our backs in the sparkling sea, or tread water and chat. We talk about families and friends, books and our work. Then, silent for a moment, we gaze at the beauty of it all. Later in the day there will be boats bobbing past, music and the shouts of holidaymakers, but early in the morning we have the place to ourselves. It is tempting to stay in the lapping water. But already I can feel Catherine is thinking of the trees she is painting and the slowed pace of the grove.

SEPTEMBER
Japanese anemone

THE BULBS HAVE arrived. The tulips are now in string bags, hanging on the hooks in the little house at the end of the garden, so the mice can't reach them. I won't plant them until the end of November and I have occasionally put them in as late as January and they were fine. Delaying the planting until the earth has cooled is meant to prevent tulip fire, the fungal disease which lasts in the soil and is terminal for tulips. Getting tulip fire means giving up growing them for several years. Wet springs usually bring it on, so the future may mean only growing tulips in pots, in clean soil, or in grass where tulip fire is rarely a problem. I like some in the flower beds – guiltily, because buying bulbs annually is not sustainable – but I can't resist a few tulips. I can justify growing hyacinths in pots for the house, because I plant the bulbs in the orchard when they are over. They come back as delicate ghosts of their glamorous portly selves, and I almost like them better as wildlings. Roman hyacinths are what I really want, but they are nearly impossible to source. They are slender multi-stemmed flowers, more like a bluebell and much airier than the familiar top-heavy hyacinth. The multiflora 'Anastasia' is a fair substitute and there are pink and white multifloras which are a good imitation of the true Roman hyacinth. Ordering smaller bulbs just means that the flowers will be smaller, which suits me. I can't give up plum-coloured 'Woodstock', which go in the pots outside the greenhouse, but they will have to wait for the 'Paton's Unique' pelargoniums to die down first.

If they are going to be out before Christmas, what can't wait are the narcissi 'Paper White'. Officially they take two months, but it's often longer. I like staggering them, so some bulbs will wait a few weeks. The little north-facing shelter with a glass roof, where I used to keep the auriculas, is now going to be for bulbs and alpines. Autumn used to be the moment for repotting the auriculas (May is when the real pros do it, but there are too many May tasks). There are still a few not exactly thriving auriculas on the top shelf. I had a long love affair with these primroses with painted faces, but it's over now. They became just too tricky and demanding. They regularly got root aphis and however much I washed the roots with neem, and scrubbed the pots and filled them with clean soil, the auriculas languished. So, in future, I'm only going to grow a few. The bottom shelf is dark at the back. It's a good place for hyacinths and the 'Paper White' to start growing roots. Once the narcissi's shoots emerge, I put them on a higher shelf in the light. Not in the house, which is too warm, but sometimes the greenhouse if I want to speed them up. This year I'm going to try giving one pot of the narcissi some dilute vodka every time they are watered. That's one part vodka to seven of water. Apparently, it stunts their growth and stops them flopping. Normally I use hazel twigs to support the flop, weaving the twigs together at the top to make an upside-down basket. The flowers look lovely growing through a hazel cage, but I want

SEPTEMBER

to see what a vodka-drinking 'Paper White' can do to support itself.

Bulbs in bowls need a lot of compost. I am mean, because they don't need fresh stuff, which is usually Melcourt and that doesn't come cheap. Bulbs in pots need no nutrients, so most years I use the spent tomato compost with added grit. Sometimes I dig some soil from the back of a bed and just add grit to that. I have grown hyacinths in water and seen narcissi bulbs emerging from a bed of stones, but soil looks friendlier. I must get some moss off the top of the field walls over the road. There are great jade cushions of it on the north side of the wall and you hardly need much to tuck in round the stems of the growing narcissi.

It's going to be lean pickings for flowers once the first frosts come, so bulbs in pots and pelargoniums, with bits of this and that for the kitchen table, is what will have to do. I try never to buy flowers, not even from the lovely shop next door. Although if a bunch of anemones appears in the dark days I might succumb. I planted some in the allotment yesterday but without a cloche, because I need them over the chicory, so there will be no anemones until spring. Pip of course has early scarlet beauties in the frame. I might try a pot in my frame here, but there is never much space to spare.

Mark and Amanda, younger neighbours in the valley two miles from here, invite me to an open evening at an organic vegetable garden nearby. I have to write about three best veg gardens for *Country Life* next year, so I wonder if this might be one of them, because I have been hearing rumours about the organic project in a huge derelict walled garden for a few years now. The owner is a fiercely committed campaigner for the organic movement, and she has invited farmers, gardeners and neighbours to a generous amount of food and drink. We listen to an impassioned welcoming speech and are introduced to Will – no socks, no shoes – who runs the garden with a team of volunteers and will take us on a tour to see progress made this year. It's a huge space, with brambles menacing the edges, waiting to advance, and Will says that couch grass is unbeatable but that slowly they are winning back the ground. He tells us it's hard and expensive to find organic manure locally, and the recycled council waste which they can get tends to be full of discarded plastic. There is a crop of celeriac which are almost as big as footballs and strong brassicas which aren't netted. I ask what they do about caterpillars. 'Despair,' Will says. To a critical old-fashioned gardener like me, so much about-to-seed groundsel and couch grass everywhere looks daunting. But new wave regenerative gardeners believe it's good to keep all the earth covered, and that weeds are better than nothing. Charles Dowding, who has inspired so many of us to grow better vegetables organically, does discourage weeds and allows bare earth. 'But that's so 2015,' a young gardener tells me.

When I get home, I look up weeds and vegetables and find several YouTube videos enthusing on 'why you should grow weeds with your veggies'. These permaculture experts are vague, but it seems to be true that cover crops between vegetables could act as a living mulch and will increase good bacteria in the soil. Clovers and legumes are the most useful, because their roots fix nitrogen in the soil. The snags are, that leaving no bare earth offers more places for slugs to hide, and that if you allow the cover crops to seed then you could lose control of the chosen weeds. I have just sown a row of green manure in a bare patch at the allotment and will hoe that off in the spring before it flowers. I think I'm not quite ready to tolerate deliberate weeds between vegetables. It's yet another example of how fast gardening techniques are changing.

September is often slanting sun and plenty of flowers. But autumn has arrived earlier than usual, bringing wild winds and far too much rain, so the garden already seems to be winding down, although the end of summer never feels final to me, or, I think, to anyone who lives with plants. Planting bulbs and sowing the quaking grass seeds, which my neighbour Trudie gave me, suggests more of a beginning than an end. While the garden sleeps, there is always the thought that next season will be even better; that there will be another chance. So, I'm planning for the year ahead.

The big border has looked good all summer. It has more shrubs in it than the middle one, which has made me think I should add some substance to the underperforming, misshapen triangle of the smaller bed. On the wide base of the triangle, next to the tiny lawn, there is an old apple tree with outspread branches. Perhaps it was once espaliered. Hot and dry on the south side of the tree and shady halfway down the bed to the north, this part of the garden has never quite worked. At the top end, next to the greenhouse, the rose 'Scarlet Fire' tangles with fennel and hollyhocks. I chose tall plants deliberately, to mask the narrowness of the bed. *Thalictrum* 'Elin' is a lofty meadow rue with glaucous leaves and, in late summer, there are several plants of the giant blue *Salvia patens*. At the shadier end now are a few intense blue monkshoods, and under the tree some pink Japanese anemones which are regularly removed, but, like all Japanese anemones, impossible to eradicate. They were here when we came and when I thought they were beaten, I planted snowdrops which have done well, so now digging out anemones means damaging snowdrops.

The theory of this bed and the even narrower one opposite, under the wall, is that on either side of the path running up to the greenhouse it should feel like a tall, traditional flower garden. So there are roses, peonies, irises, delphiniums, asters, dahlias, salvias, all the familiars, with plenty of green and something to see all year. The big bed has less variety; there are species roses like *glauca* and *xanthina*, with bulky shrubs in scale with the inherited silver pear, and purple berberis on the boundary with Bob. This style of planting, what Laura calls 'contrastifolia', takes me back to where I started gardening, when silver and purple shrubs were admired by my mother-in-law Catherine. Five years ago James Wong wrote in the *Guardian* that shrubs were unfashionable but no longer deserved to be sidelined. Since then, avant-gardeners have been upping their shrub quota. Fergus is always a step ahead of the rest of us at Great Dixter. He now champions conifers, although it may be a while before other gardeners follow his lead. But here I am, thinking that what the wonky triangle bed needs is a shrub to pull it together and link it to the more successful bed opposite.

In my own garden I can take risks; if something doesn't work, I can change it. A shrub – purple, almost black, with filigree leaves – is going in this afternoon. It's a classy form of the common elder, *Sambucus nigra* f. *porphyrophylla* 'Eva', which will need radical pruning to keep it in bounds, but I'm looking forward to the pink flowers against deep purple leaves, to autumn colour and to berries for the birds. If the Japanese anemones insist on staying, they will look all the better in company with the dark leaves, echoing the plum-coloured foliage of dahlia 'Magenta Star' and the grey leaves of *Rosa glauca* in the bigger bed across the grass path.

The best border plants now are the Michaelmas daisies. Not the stiff-stemmed New England ones, which tend to lose their lower leaves and get mildew in dry places, but the smaller-flowered bushier plants. 'Little Carlow' is most people's favourite, but I like the airier *turbinellus* better, and a new to me pink form called 'Vasterival', with darker stems. Much lower than these is *Symphyotrichum lateriflorum* (*Aster horizontalis*). I used to grow Aster × *frikartii* 'Monch' and can't think why I stopped because it flowers for months. But it is a bit more purple than blue and has larger flowers than my favourites, so maybe that's why I don't.

It's an outing to Chippy Flower Farm in the high north Cotswolds, for an article for *Country Life* magazine. Tif and Anna, two friends who started the enterprise, are welcoming and their enthusiasm is infectious. In three years they have turned an acre of field into a thriving business. They do flowers for weddings and funerals as well as pick-your-own. I never thought of using raspberry leaves in an arrangement, but they do, and when I look properly at leaves rather than berries, I understand why. They use herbs, too, especially mint for scent, and are about to run a day workshop on dried flowers. They like 'dead beauty' and want people to enjoy the experience of seeing real flowers grown locally.

When I am leaving, a man walks up to me and asks if I am Mary Keen. He says, 'I cut the topiary at your old garden', and pulls out his telephone to show me a picture of the enormous bulge of yew above the old flower garden. It's immaculately clipped, but below it are no longer flowers, only lawn. I say, 'What happened to the garden?' and he says, 'Oh, that was too much work. They've got children.' I know they have children, the people who bought our last house are a delightful family. I know, too, that neither a house nor a garden are for ever, but I can't help feeling sad. I used to call that bit of the garden 'sub Sissinghurst', because, like the cottage garden there, it had a large copper at the centre of four beds, guarded not by Irish yews but by the fastigiate box 'Graham Blandy'. The beds were uneven sizes, divided by a diagonal St Andrew's cross and packed with flowers. I could stand under the old holly tree above to look down on the plants or walk down the path that formed one side and then wander through the beds, surrounded by colour and scent. Thinking about that vanished place, I realise that immersion is what I always want in a garden. I want to be able to lose myself among high plants on either side of me. And I console myself with the thought that I can still do that here, in this tiny plot. I can set off through the two borders on either side of the gravel path that leads to the greenhouse, going in at one open door and out at the far end. Making the greenhouse a passage means that I can walk an unbroken loop around the garden. When I reach the top, I can carry on down the mown path through the meadowy orchard, until I am between the big bed with shrubs and the other side of the bed which I have already passed on my way to the greenhouse. Then it's a few steps across the lawn and back down the paved path with arches to the courtyard, where plants grow right up to the kitchen windows. On every step of the way – three hundred at the last count – I am cocooned in a world of plants.

Alice has come to garden this afternoon, and she is going to dig out the tall *Althaea* with tiny pink flowers from the apple tree bed and help me move it to the other side of the path. It's too big where it is, almost as tall as the apple. Its outsize scale will work better if it moves across the path, where it can give the blue Adirondack chairs a more secret corner. The *Althaea* seeds itself obligingly, but it can also be divided, and I like the idea of a large plant next summer. Division turns out to be quite an undertaking – there is a taproot the size of a large parsnip – but we wrangle some shoots with smaller roots away from the parsnip and plant those, as well as a few more in pots. Another self-seeder is the favourite *Euphorbia characias* subsp. *wulfenii*, and I have a good-sized potted one ready to plant in the gap left by the mallow. This, with its lime-green flowers and grey-blue leaves, will be perfect, and I doubt it will get higher than a metre and a half. There are several of the same euphorbias in the garden already, but there is nothing wrong with repeating a good plant. When I find useful seedlings of the spurges or the crimson-flowered silver-leafed campion, I pot them up and put them in the frame, so there is always a plant to fill a gap. I don't like being without my favourite wild *Dianthus carthusianorum*, so there are reserves of that too. Like all pinks, it doesn't live long, so there are plenty of rooted cuttings which were taken in June, and this means I can replace any plants which have passed their best. There is an added reason for keeping my own stock going: the dianthus is a much taller form than the one that is generally on offer. Plants make good presents, so having a surplus is never a problem.

September is prime time for building up a reserve of plants. All the half-hardy salvias are on the list and most of them are now in the greenhouse, five sitting round the edges of a pot. Favourite pelargoniums will soon be inside, some as big plants cut back and others as cuttings, which won't be potted on until spring. I'm going to save more dahlias indoors this year. In the past I have left them in the ground, covered by a deep mulch. It is generally accepted that in our warmer climate, dahlias survive the winter outside. When they fail to emerge in spring, I doubt it is cold that has killed them; it's slugs, eating the emerging shoots. I think the answer may be to grow them in pots until they are less tasty and then put them out. Slugs love rain and young tender greenery, and this year my dahlias were far from their usual abundant selves.

I have been less able to get to London to see exhibitions this year, so am glad to catch the 'Gardening Bohemia' show at the Garden Museum and am struck by how much their gardens meant in wartime to these women, how Vita Sackville-West could write in her long poem, *The Garden*:

> Small pleasures must correct great tragedies,
> Therefore of gardens in the midst of war
> I boldly tell.

There are Vita's beautiful tools with extra-long handles and the initials V.S.W burnt into the wood. I covet her scaled-down version of a mattock. I never see anything like that now and I am a tool nerd. I regularly use a small hand weapon shaped like a mattock for weeding, better than the standard onion hoe for leeks and onions, but which doesn't have enough leverage for hooking out difficult weeds like brambles. You need a longer handle for that. I suspect Vita's mattock would be a treat to use.

Piers and Louise of Garden and Wood and the triple-decker bee orchids might know if Vita's tool was specially made for her. Their business is vintage gardening. They find, mend and sell old garden things, sending sprinklers to Ernie in America, who collects irrigators as well as snowdrops. It is their terracotta pots which I cannot resist, but I must remember to ask about the elegant mattock next time I see them.

The other Bohemians at the garden museum are Vanessa Bell, describing and painting a 'dithering blaze of flowers', and her sister Virginia Woolf, writing of summer in the orchard, seeing 'the very topmost leaves of the apple tree, flat like little fish against the blue'. The fourth gardener is the hostess Ottoline Morrell, but did she physically garden like the other three? Garsington became a refuge for anyone who was entertained there, but Monk's House, Charleston and Sissinghurst were sanctuaries, and that and the way their owners worked in their gardens makes me think that this was close to our recent experience of lockdown. Those of us with gardens then were lucky, and when I garden now I like to think that after a day among the flower beds the elegant Virginia Woolf could, like me, be 'stiff and scratched all over, with chocolate Earth in our nails'.

Tim, the tree surgeon in the village, removes the metre-wide lonicera hedge at the edge of the allotment plot today. Its roots stretched another two metres underground, so plants in that area have always been stunted. The last plotters, who are good growers, had advised keeping it because it was a useful windbreak, but lonicera needs cutting three times a year, and it's ground-gobbling tendencies have been getting me down. Tim does an expert job, but the roots are still waiting for the stump grinder to make sure not a shred of hedge remains. It's been so wet that taking heavy machinery over grass is risky and anti-social, but I'm hoping there will be a dry patch soon. Once that is done, I might plant step-over apples instead of the hedge. I've never grown them, but would like to try.

I go to Petersham Nurseries. Pip and I did some work on the garden there years ago, so when Lucy Bellamy asked me to come and talk about her new bulb book I couldn't resist. We chat about sustainable bulbs and choosing the best kinds and when to plant. When I come out rather strongly about getting narcissi planted in September, Thomas, from Petersham, wonders if that rule no longer matters, because we should be guided by waiting for cooler temperatures. Interesting. Francesco gives us a tour of the garden, where now the topiary from Pistoia is massive, dwarfing the Gormley statue still rooted to the spot. Long ago, for another client, I spent an afternoon with the sculptor himself, wading about over heavy clay, in order to decide where his statue should stand. We clomped over the sticky ground until he came to rest with his arms hanging stiffly down, but not touching his sides, so that he was his own breathing statue, standing in platters of clay. If the rusted Gormley has stood the test of time at Petersham, the long borders are a bit changed. How could they not be, it must be all of twenty years.

OCTOBER
Pears

A UTUMN IS WHEN modern perennial gardens, like those of the great plantsman Piet Oudolf, come into their own. Pip's theory is that all plants look good until they flower, then they collapse, so it might make sense to have a higher proportion of autumn to spring performers. Not always true, as peonies, especially the intersectional ones, have lovely leaves all summer, and some irises, particularly *pallida*, stay handsome and upright after flowering. I try not to cut down any of the bearded irises in the approved fashion, unless the leaves start to brown. This year, because of overwhelming rain, has been less good for that tactic, but in a dry summer the blue-grey fans of sword-shaped leaves can be as beautiful as any flower. *Iris sibirica* has large seed pods, so that is worth planting for two seasons of interest. And I like the leaves of *Baptisia* after it flowers. But Pip is probably right, he usually is, there isn't much that looks good after a spring bloom.

Most of the Michaelmas daisies are still out, but flowers of the month are chrysanthemums and they should see me through until November. I grow them over the road at the allotment and I've just brought back a huge bunch of favourite 'Emperor of China' and 'E. H. Wilson'. I've taken against 'Bretforton Road', which is the only other one that has survived. With shocking pink, shell pink and salmon pink I'm fine, but pink on the mauve side is my least favourite colour. I didn't order more varieties this year to replace 'Allouise' and 'Tula Green', which are only half-hardy. 'Tula' is a spidery-petalled orange and has survived in a pot in the greenhouse with minimal attention. I'll lift the 'Dixter Orange' in the garden, even though it's meant to be hardy, because otherwise every shoot is eaten as it appears in spring. Maybe next year I will be kinder to chrysanths, I've been a bit lax this season. Arthur P. says they need more water than dahlias. *Gardens Illustrated*, out already, has a good article about top hardy types. I am tempted by 'Chelsea Physic' and will try 'Ruby Mound' again. 'Burnt Orange' has spiky spoon-tip petals. All these are on offer from Norwell Nurseries in Nottinghamshire, so if they survive there, they ought to be fine further south. I just telephoned Norwell to ask about ordering, because it didn't seem possible online. 'Post the order form with a cheque,' said a voice at the end of the phone. A cheque? Does anyone have a chequebook these days? But he sounded friendly, so I asked him why my 'E. H. Wilson' chrysanths had shrivelled brown leaves this year. 'Probably eelworm,' he says. 'It's always worse in a wet year, and old varieties tend to get it, but they usually survive.' I've been stripping the leaves off the stems for picking, there are some picked flowers on the kitchen table now; still scented, they are perhaps a little smaller than normal. Looking up 'eelworm' – I hate looking up pests and diseases, but sometimes you must – the RHS says you can soak the dormant roots in hot water for five minutes. But the temperature has to be exactly 46 degrees Celsius, followed by a cold plunge, which sounds difficult to arrange. I'm tempted to try neem, the best permitted pest control I know. And to order some more 'E. H. Wilson' for next spring in case that fails.

It's a Jane day and we dig up the rest of the dahlias. I know they usually survive winter under a three-inch mulch, but I'm hoping that slugs will not destroy them if I can put them out as stronger plants next May. Arthur says bring them into the house to dry the tubers off, but I've always stored them upside down under the greenhouse bench and then potted them up in dry compost to wait until early March for their first drink of water.

In the gaps left by dahlias and other half-hardies, we plant out the apricot foxgloves, which I sowed in June. Last year I put them in the allotment to grow, but they didn't, so this year they were potted on for a spell in the frame until there was room to set them out in the beds. There are far too many self-sown honesties and primroses in places where foxgloves should be, so the intruders get a cull, while Jane goes another round with the pink Japanese anemone under the apple tree. It seems contrary to fetch the rare white form from the passage outside the garden that leads to Back Lane. I've been admiring it there and I think it deserves a place nearer home, even though *hupehensis* anemones are the most invasive of plants. Jamie, the great plant hunter, collected it in Szechuan and brought it as a moving-in present when he and Tania came to lunch five years ago. My private reaction then was, no thanks, it will be as troublesome as the ones already in the garden, which I can never dislodge, so I planted it beyond the pale in the guerrilla garden passage. Moral: when a serious plantsman gives you a present, it should always be yes please. It's a lovely single white form, so it has now been welcomed into the wall border, next to the myrtle near the greenhouse, where it can be regularly admired.

The guerrilla garden has got a bit out of control this year. Partly my fault for being away and partly the fault of the montbretia, which has bulked up so much that its leaves are forever flopping on the path. I've pinned the greenery back with wire hoops, but the plants need dividing and reducing. This common form of crocosmia, which grows wild in the West Country, is a star plant. Its leaves look fresh and handsome in spring, then orange flowers follow, and finally there are elegant curls of stem set with orange seeds. Some of the seed heads are on the kitchen table now, with a few 'Dixter Orange' chrysanths and the first sprays of winter jasmine.

Periwinkle is also on the rampage in the passage, and I see Simon has dug out some and replaced it with bearded irises and blue knapweed. As it's a joint area I can't interfere, but ideally this narrow path should look more like a country lane than a garden.

Dismay in the village, because the shop of shops is going to be sold. The sisters who own it are giving up. They have made a great success of it since the other one over the road burnt down, but it must be hard work keeping the antique shelves stocked and remembering which mahogany drawer has fuses and where to find the Sellotape. I hope somebody equally sympathetic buys it and will carry on selling eco refills, as well as Charlie's home-grown veg, olive oil and best bakery, on top of all the basics. Apparently, lots of people have looked, but because it's listed and the interior can't be changed, there are no takers so far. It's been good and profitable as it is, but potential buyers can't seem to believe that anything so unlike a modern supermarket could work as well. Most customers help themselves, because there is room to get behind the counters to reach for tea or coffee on one side and veg and fruit on the other, so it's self-service for everything except cheese and there is always someone at that end of the shop.

There are deer at the allotments; they seem to be coming in from the west. Tony reported them on his plot, and I selfishly hoped they would find enough to eat before they reached my crops further east. But the endives have been badly chewed, so now everything must be netted or cloched. Ellie has deer in her Bristol allotment, and occasionally threatens to give up growing because the depredations there are so disheartening. She also reports many attacks of ticks, and ticks equal Lyme disease. I suppose we too must keep a close watch for them, which means searching for specks of black the size of a poppy seed after every visit. Wearing wellingtons, with trousers tucked in, is meant to prevent ticks reaching any skin, but I prefer leather boots, and you can't tuck trousers into those. Bicycle clips might work. The deer problem makes me think more seriously about investing in a polytunnel for winter salads and later tomatoes.

This year, when plants are past their best, I'm going to try the new-fangled 'chop and drop' system in the flower beds. The stems get cut into small pieces and are left lying on the bed, which saves all that carrying to the compost heap and barrowing back as rotted matter in spring. I need to be careful to keep away from emerging snowdrops, scillas, primroses, forget-me-nots and co. I will add small leaves to the back of the beds under roses and around things like campanulas. The magnolia leaves are too leathery to use. The end result might look a bit messy, but the RHS practises 'chop and drop' and says it saves time and increases biodiversity. I'm hoping it deters cats too, but I worry that the self-seeders will be thwarted.

My friend Sally up the road says cyclamen do well in the village. Not for me they don't. When we came, I bought an extravagant number of brightest pink *coum* tubers for the winter months. There should be some under the apple tree and some in the beds outside the kitchen window. There are a few, but not as many as I would like. If you look up how to grow them, most advice suggests siting them in shade. But at our last garden they seeded themselves into a sunny dry slope at the end of the lawn, and I suspect what they want above all is good drainage. Under trees in dry shade suits them, so the apple tree ones have survived. The beds outside the kitchen window where I planted plenty are behind the groups of pots in the courtyard. Perhaps the regular watering means that the beds collect the run-off from the pots. I must be more careful next year. The autumn-flowering *C. hederifolium* are doing better. They are increasing in grass under the wall at the far end of the garden. This is the patch in front of the miniature stone house where snowdrops and aconites grow. We mow it through the summer, but that doesn't seem to bother the cyclamen, because the end of the mowing season roughly coincides with their arrival. My favourite cyclamen is the spring-flowering *repandum*. It's scented and a bit larger than the other two and even trickier to grow. Damp shade and protection from frost suits *repandum*. 'Moist but well drained' is the worst condition to arrange, but it is what so many desirable plants seem to need. West Country gardens often have sheets of this Mediterranean cyclamen. I wish I did, but helpful ants have moved some seeds a bit away from the parent plant and a few baby plants have appeared, so maybe in time there might be more.

Shocking pink is a tonic in the grey days of autumn, so I am wondering where I might plant hardy nerines. In the last garden there was a narrow bed under the south wall of the schoolroom next to the path. Here, I've used up too much of the only hot wall at the top of the garden by putting a seat under the pergola covered in rose 'Francis E. Lester' and the wrong form of *Lonicera etrusca*. To the left as you face the seat is shaded by the crab apple over the green bin zone. To the right there is a wintersweet with a few pale pink amarines, a glamorous nerine–amaryllis cross. They've been hardy, but I've never tried straight pink nerines in this garden, except a few of the more tender forms in pots. My Irish friend Christopher has them growing in grass, so perhaps I could try some in the grass in front of the seat as well as in the paving, where the Mexican daisy has settled. Better still might be the tiny bed in front of the scaled-down cottage at the far end of the garden. But only one side gets enough sun, which would mean the beds would no longer vaguely match. Now they have small bright flowers, geums, forget-me-nots and 'Tête-à-Tête' daffs, as well as the rose 'Madge' on either side of the door. Flowering non-stop, this small, porcelain-pink musk rose was a favourite of Christopher Lloyd, that champion of exotic clash. I like this reminder of the great man and of the pleasure of staying at Great Dixter. There was always a rich meal in the evening, cooked by Christo, followed by sitting in the solar, with a dachshund lapping the dregs from the coffee cup. The bedroom was reached by a landing with smooth, wide floorboards, and after breakfast there would be a tour of the garden. A notebook was encouraged. Christo used to tell visitors who asked for the name of a plant that he would not answer their question unless they could write it down. I learned to bring a rubber notebook for writing in the rain.

Every October Matthew Jebb, director of the Irish Botanic Garden at Glasnevin, organises a working party to Lambay, the small island in Dublin Bay where the only buildings were either restored or built by Lutyens. At its heart is a romantic castle, surrounded by a wood which is enclosed by a circular wall. The place can hardly have changed since Lutyens time. We stay in the White House, which was built in the 1930s for the two daughters and ten grandchildren of Cecil and Maud Baring, who bought the island in 1904. It's still in the hands of their descendants, who are determined to keep Lambay going, but it's hard to know how such a remote place can be made to pay for its upkeep. Tides and rough weather mean that getting to and from the island can be unpredictable. There is no shop on Lambay, so Matthew's wife Serena organises meals and provisions for a week, for as many as ten. Everything is brought over by boat. Serena cooks delicious meals for the garden team and the house is warm with a luxurious number of bathrooms, all heated off-grid, so it's more like a holiday than hard labour. There is even time for a tour of the castle and, for some, a swim in the sea, which this year I resisted.

The bones of a garden laid out by Jekyll are still there around the castle, but with no regular skilled help it's a struggle to keep plants in bounds. Last time we concentrated on the wall-trained fruit trees in the walled garden; this time it is roses that get most of my attention. Kew-trained Beth does some impressive weeding and her artist daughter Tess, who has never gardened before, soon learns how to pin roses into loops and curls on the walls. After I show her a picture of Jenny Barnes's curlicues, she is longing to create more ambitious patterns, but we stick to severe and simple, because it keeps for longer. While we sort out the roses, Christopher, Michael and Kirsty fight their way through overgrown borders and Louis barrows away mountains of pruning cuttings and weeds. Everything grows so fast on Lambay that the garden gets out of control within months. 'Cut it to the ground,' Matthew says of a flowering fuchsia. So, we do, even though it is still beautiful. When Beth's bed in the West Court is clear, we suggest adding some crocosmia to cover the empty ground. There is plenty but Matthew is fierce and says, 'Nothing but what grows wild on the island, or what Miss Jekyll would have planted.' So, we don't. In order to keep us in awe of Jekyll, Matthew sends everyone an email with a coloured plan of one of her borders, next to a picture of Turner's *Fighting Temeraire*, to show us how much she was influenced by great art. I'm impressed, but perhaps more by Matthew's ability to colour her black-and-white scheme and make the connection with Turner, than by Miss J's sense of colour. I know that once her eyesight started to fail she had to give up embroidery in favour of gardening, so I wonder if she could see the Turner painting clearly enough to copy the colour scheme.

If crocosmia gets the thumbs down, the suggestion that the bed that Christopher has weeded should be grassed in future does get approval. This seems sensible. Grass is regularly mown, but it is unrealistic to think that there will ever be time or skill to weed a border once the working party leaves. Christopher has two gardens, one in Kilkenny and one in Burford, which he seems to manage in the intervals of a full-time job advising on the conservation and decoration of most important Irish houses. Practical as well as creative, he has travelled with his strimmer, so he trims all the edges of grass plats and paths, and the design is so strong it doesn't matter that there are no flowers. This year it doesn't rain, and on the last day we walk to the little beach where there are seals and cowrie shells.

It's been a year of learning about new ways of gardening, because I want to understand how we can adapt to changing times and extreme weather patterns. The last lesson is in north Devon. I was impressed when I heard Joshua Sparkes give a talk when he was head gardener at Forde Abbey. He was clearly someone to watch, but after he left Forde, he began a whole new project in a seventeen-acre field at Woolsery, growing edible crops commercially, which is probably going to change the way everyone gardens in future years. Josh's system owes something to Japan, something to permaculture, something to regenerative agriculture and quite a lot to his own intuition and experience. It's very complex. Interlocking root systems seem to provide all the nourishment needed to grow crops good enough for a gourmet pub with Michelin recognition, as well as for a destination village shop. No composting, no digging and no hoeing means that three men can manage seventeen acres between them and Josh is often occupied with teaching or lecturing, so call that not quite three. The National Trust used to say that it was out of the question for one man to cultivate more than three intensive acres, but here the ratio is almost double that. Josh says that biodiversity still matters more to him than productivity. But increased

OCTOBER

biodiversity means increased productivity. After five years he finds that all the worst weeds have disappeared, that cabbages don't need netting because natural predators will keep them clean and, best of all, the beetle banks at regular intervals all over the field have eliminated slugs. As he shows us round, Ellie and I are open-mouthed. On a wet winter's day, it still looks very beautiful and alive, with plenty of flowers as well as vegetables. Next year's two-day course is already sold out. Bother.

Today, at the start of the last week in October, Fergus on his Instagram marks the end of another gardening season at Great Dixter, adding, 'Our gardening year starts now. Many exciting plans and fantasies already.' Another chance holds hope for all gardeners, and this morning the sun is shining as I come to the end of the last month in a year of writing about what keeps me hoping. Has there been a single hour when I haven't thought about, talked about, read about or learned about growing plants on the rare days when I have not managed to be outside pottering? Pottering won't stop in winter. And in winter there is also peering as I search daily for signs of returning life. I will watch for the first snowdrop tips to emerge, and wait for blossom on the autumn-flowering cherry. I will stand and stare at the fur-wrapped magnolia buds and wonder how they could possibly enclose next year's enormous flowers. There will be the greenhouse to open every morning and shut every evening. A few late roses on the tender courtyard climbers will surprise me. And, once I step indoors, I will remember the bulbs I have forgotten to order, or the seeds which ought to be sown, and I will start to dream of another gardening season, which will be quite different from the one just past.

A List of the Plants I Currently Grow

Personal observations on all the plants grown in
my garden, which is on well-drained limey soil in
the west of England at an altitude of 160 metres.

Occasionally I suggest plants I have grown for others, or which I wish I had room for here, although I may have missed some and others may have dwindled since the time of writing. Having made these lists, I realise I cannot resist growing far too many different things and that scale and structure, which do matter, are less of a feature for me than they might be in the gardens I make for others. The atmosphere of a garden is always my priority, and peering at small plants is what I enjoy doing at home. Current thinking about biodiversity encourages us to grow as many different plants as possible, so this may be the way forward for us all.

Of watering, I am still occasionally guilty, but I try to reduce this, which may mean giving up plants like phlox. I do not use pesticides apart from a very few 'organic' slug pellets. I must stop this, as, in spite of their claim to be safe, it is now known that they are damaging to wildlife. Even the best biodiverse gardeners admit that they tear their hair out for two to three years before finding that the balance of nature is restored after abandoning all chemicals, and I may not have that much time. I do not use weedkillers and the only fertiliser I use is a seaweed mixture from a bottle: a very good organic gardener once told me that greenfly cannot bear the smell of it.

Winter Interest: December, January, February

Trees and Shrubs

Acacia dealbata: mimosa, yellow flowers after Christmas in mild winters. Worth growing for its ferny leaves. Technically unsuitable for Cotswold conditions, but I like a challenge, and after a hard winter it should grow back from the ground.

Azara microphylla: vanilla-scented tiny flowers, very straggly evergreen growing in shade here but would be denser with more light than I give it.

Buxus **'Graham Blandy'**: the fastigiate box, makes a slender exclamation mark.

Buxus sempervirens: in cloudy unclipped shapes make good winter punctuation and are less prone to blight grown freely.

Chaenomeles **'Rowallane'**: blood-red flowers on old wood, best wall-trained.

Chimonanthus fragrans: wintersweet. One twig will scent a room. Needs sun and not attractive out of season.

Cornus alba **'Elegantissima'**: a sentimental choice because it grew in our earliest garden on the banks of the Lambourn. I cut it back every spring so that it produces a bunch of red stems in time for winter. In summer the gently green and white leaves attract some adverse comments from gardeners who despise variegation, but it is here to stay.

Coronilla valentina **subsp.** *glauca* **'Citrina'**: pale yellow pea flowers all winter, but this is a short-lived shrub for a sunny place. Needs regular clipping to keep it from flopping too much.

Daphne bholua **'Jacqueline Postill'**: tall, scented, usually evergreen (but not for me), it suckers if happy and I pick sprigs all winter. A tiny twig will scent a room.

Edgeworthia chrysantha: scented flowers in February like little yellow parasols

Hamamelis mollis **'Pallida'**: witch hazel with pale yellow flowers and wonderful scent. I grow it in a pot, where so far it is fine, but I don't let it dry out in summer.

Ilex aquifolium: Christmas holly. There was a tiny seedling growing in the bed near the greenhouse, which I moved, did nothing to further its growth and after six years it is nine feet tall. Memo to self, this is the best way to establish trees. I will probably clip it when it gets taller, so as not to shade Bob's garden next door.

Jasminum nudiflorum: yellow winter jasmine. Old shoots which have flowered should be cut back in spring.

Laurus nobilis **f.** *angustifolia*: the narrow-leafed form of bay is much easier to manage in small spaces and it tastes exactly the same as the familiar version.

Luma apiculata sent in error for *Myrtus communis*, but happy against the wall next to the greenhouse. Its bark is interesting but its leaves are a bit drab.

Melianthus major: glaucous leaves unless winter is extra cruel, but it comes back fast from ground level. In hot summers a strange tawny flower may emerge.

Myrtus communis **subsp.** *tarentina*: hardyish small-leaved myrtle which can be clipped into shape.

Nandina domestica: the straight unimproved form is lovely all year. White flowers in summer followed by scarlet berries, red-tinged leaves in autumn and it's evergreen. Happy in a pot in the darkest corner of the courtyard here, but probably even better in the ground in some sun.

Pittosporum tenuifolium: already in the garden. It makes a large evergreen tree with tiny scented almost black flowers in spring. I would probably not have chosen to plant it myself. *Pittosporum tobira* is smarter and glossier with scented white flowers for mild gardens.

Prunus lusitanica: best kind of Portugal laurel with handsome leaves and white flowers later.

Prunus mume **'Beni-chidori'**: shocking-pink blossom in February. It flowers on new shoots so needs pruning after the flowers fade. Can be wall-trained.

Prunus × subhirtella **'Autumnalis'**: winter-flowering cherry, with pink or white blossom. Mine, *P.* 'Autumnalis Rosea', is a tree, but it can also be grown as a shrub.

Ribes laurifolium: a flowering currant with hanging green tassels. It's one of the prettiest and most intriguing plants in the garden, but it grows very slowly. Lovely hanging over a low wall.

Salix **'Mount Aso'**: a pussy willow with pink catkins. I grow it at the allotment for picking.

A LIST OF THE PLANTS I CURRENTLY GROW

Salix **'Nancy Saunders'**: has whippy red stems for wreath-making and little catkins in spring if you leave it to grow, but I cut it back in late March to keep it small and airy.

Sarcococca confusa: honey-scented Christmas box, makes a huge plant. Best in shade, even dry shade.

Sarcococca hookeriana **'Winter Gem'**: a more delicate version of Christmas box.

Viburnum × *bodnantense* **'Dawn'**: dependable scented winter shrub.

Viburnum **'Charles Lamont'**: said to be the best form of this scented winter shrub. I'm still not convinced, but it is pinker than 'Dawn'. Both are happy in sun or shade. They need no pruning but can start to look craggy so I do occasionally cut a few old branches right down.

I'd like to have room for *Garrya elliptica* 'James Roof', with particularly long catkins, and almost any of the mahonias for their yellow lily of the valley scented flowers. *Mahonia* 'Winter Sun' probably, if I have to choose. *Cornus mas* is desirable and quite happy on dry soil, which books will deny. All these shrubs are invaluable in winter, but a bit dull in summer in a small garden.

Perennials and Bulbs

Adiantum capillus-veneris: maidenhair fern, which grows in the passage outside the back door in an old copper. Not as tender as people say. I forgot to water mine in the recent exceptionally dry spring, but after adding a little washing-up liquid to the can, the fern revived within days. If a pot is very dry, the addition of soap to the water helps it soak into the compost.

Adiantum venustum: another maidenhair fern in the ground in shade. Very pretty, hardier and much smaller than *A. capillus-veneris*.

Asplenium scolopendrium: the hart's tongue fern needs dead leaves removed in spring to stay glossy.

Bergenia: I only grow two of these. Silly perhaps, because they are useful. *Bergenia ciliata* is less coarse-looking than the shiny *B. cordifolia* and its giant leaves make a good summer full stop. I love the delicate not quite hardy *B. emeiensis*; I grow the white form and would like to find the pink.

Cardamine quinquefolia: has lilac flowers when the snowdrops are out. It spreads fast in shady places but being summer deciduous, it vanishes completely in summer.

Crocus tommasinianus: it is increasingly hard to find the unimproved form of this early delicate crocus. If you are lucky and can find the right thing, once it settles it will seed. There is nothing lovelier.

Cyclamen coum: the best winter tonic. I grow the pale pinks, or they grow themselves with the help of the ants who move their seeds around, at the far end of the garden among aconites and daffodils. The stronger pinks come up in patches where they want to be. I try to keep them away from the bright yellow aconites and 'Tête-à-Tête' daffs which are also in the mix, with plenty of common snowdrops.

Eranthis hyemalis: aconites with shiny yellow cups in green ruffs. Buy and move them in the green. There are different forms. Mine appear later than I would like, so I am looking forward to trying *Eranthis cilicica*, which is said to flower earlier.

A LIST OF THE PLANTS I CURRENTLY GROW

Galanthus, or snowdrops, are a cult thing, and the good ones are expensive. I used to love collecting them but it's harder to find places where they can grow undisturbed in a small garden. Some of the ones I grew in the old place have persisted, but I have lost the beautiful 'Daglingworth', which was a favourite form, and the 'yellow' 'Spindlestone Surprise' is dwindling. I plan to try 'Madeleine' instead. I've failed twice with the October-flowering *reginae-olgae*, which likes a sunnier place than most.

Galanthus **'Anglesey Abbey'** has grass green leaves.

Galanthus **'Augustus'** can be virus prone, seems happier in grass.

Galanthus **'Desdemona'**: a double and bulks up fast.

Galanthus **'Diggory'**: bell-shaped, so easy to recognise.

Galanthus **'Faringdon Double'**.

Galanthus **'Fishing Rod'**.

Galanthus **'Galatea'**: reliable.

Galanthus gracilis: has curling greyish leaves and is always said to produce promising crosses with other snowdrops.

Galanthus **'Hill Poë'**: a reliable double.

Galanthus **'Limetree'**: good spreader and early.

Galanthus **'Merlin'**: green inside. I love the idea of a green snowdrop, but they are still too expensive. I do have the virescent 'Kildare', with palest green streaks on the outer segments, but you have to peer to see them. G. 'Rosemary Burnham' or G. 'Cowhouse Green' would be more exciting, but neither are cheap.

Galanthus **'Modern Art'**.

Galanthus **'Mrs Macnamara'**: very lovely, early and easy.

Galanthus **'Natalie Garton'**: larger than most and easy. Very pleasing.

Galanthus nivalis: the wild snowdrops in early, late and double forms.

Galanthus **'Peg Sharples'**: one of the last to flower.

Galanthus plicatus **subsp.** ***byzantinus*** with broad leaves, increases faster than most.

Galanthus **'S. Arnott'**: handsome, scented, easiest of the tribe to please.

Galanthus **'Three Ships'**: out for Christmas.

Galanthus **'Trumps'**.

Hellebores are a welcome tribe as winter turns to spring. I used to like the nearly black strains but now I think that these never show up as well as the pale colours do against a background of earth. The Lenten roses *Helleborus* × *hybridus* come in all shades, but flowers with rounded, not starry petals and never double are my favourites. I particularly like the ones with dark nectaries or a picotee edge. Ashwood Nurseries have a mouthwatering selection. *H. niger*, the Christmas rose, is not as easy as the *H. orientalis* types. I think it does better in a little sun, although most advice suggests partial shade, with that well-known, impossible to follow caveat, 'moist but well drained'. The Corsican hellebore, *H. argutifolius* has apple-green flowers and handsome leaves. It likes a little shade and dislikes moving or being divided. The stinking hellebore *H. foetidus* is happy in most places. I would give it a more open slot if I had room to spare, but it hangs on in a dark corner of the sunk paved patch at the back of the extra room beyond the courtyard. All the hellebores are great self-seeders and if there is somewhere to grow the infants on, then you can see if they are worth keeping. Some of the crosses can be very murky. *H.* 'Anna's Red' is large and showy with striking marbled leaves. I grow it in a pot where it looks glamorous, never in the ground, because it can look a bit overdressed among the early fragile winter bulbs.

Irises are another late-winter pleasure, some in pots, some in the ground.

Iris histrioides **'Lady Beatrix Stanley'**: an old favourite for a sunny dry place, with perfect blue flowers.

Iris lazica from the Black Sea, looks like the Algerian iris but is less demanding to grow and manage. Its wide grass-green leaves stay tidy and it is happy in shade. Slugs love the buds of these winter beauties. I have an unnamed darker form, given to me by my Irish friend Christopher, and a paler one from Tania Compton.

Iris reticulata **'Blue Note'**: dark and fragile.

Iris reticulata **'Fabiola'**: blue-sky blue. I try others in other years.

Iris reticulata **'Katharine Hodgkin'** and **'Katharine's Gold'**: unusual and beautiful.

Iris unguicularis **'Mary Barnard'**: the Algerian iris for a hot, dry spot. It flowers on mild days all winter. Pick the buds to open indoors. The grassy leaves can look messy in summer and removing the dead ones is a fiddle. *I.* 'Walter Butt' is paler and both are faintly scented.

Narcissus **'Rijnveld's Early Sensation'**: tucked into the bed under the wall between the greenhouse and the bin corner. It's for picking in January and there are also some in the guerrilla garden passage. *N.* 'Tête-à-Tête' is almost as early. There were some here when we came. Moved and divided, they have bulked up fast in the bulb area in front of the little house.

Pachyphragma macrophyllum: evergreen large-leaved plant with white flowers like candytuft in late winter. Happy in rubbish soil and shade.

***Polypodium calirhiza* 'Sarah Lyman'**: a summer-dormant fern. Beautiful in winter.

***Polystichum setiferum* (Divisilobum group) 'Bevis'**: the best fern for glamour and all year presence.

***Polystichum setiferum* (Divisilobum group) 'Herrenhausen'**.

***Primula* 'Hall Barn Blue'**: desirable early named primrose.

Primula vulgaris: the common primrose seeds everywhere. It is a weed in the flower beds, where I leave it until later flowers need the space, but it is a welcome sight for its green leaves all winter. In mild years there are flowers too.

Pulmonarias, the lungworts with spotted leaves, flower from the end of winter into spring. There are masses. I favour the red and early or the bright blue. I am less sure of the two-tone blue pinks, but *P*. 'Diana Clare' is here and handsome. The lungworts will grow in shade or semi-sun. In a dry summer their leaves will die. Cut them back and water them to grow again.

***Pulmonaria* 'Blue Ensign'**: lacks spots on its leaves but has gentian blue flowers.

***Pulmonaria* 'Dora Bielefeld'**: small, a quick spreader with coral buds opening to pink.

***Pulmonaria* 'Leopard'**: has good leaves and coral coloured flowers.

***Pulmonaria* 'Opal'**: a cool blue, very nice indeed.

***Pulmonaria* 'Redstart'**: early but not showy.

A LIST OF THE PLANTS I CURRENTLY GROW

Pulmonaria **'Sissinghurst White'**: lovely.

Vinca difformis: milk-pale periwinkle flowers all winter and off and on all year, but it is determinedly invasive.

Indoor Plants

Begonia albopicta **'Rosea'**: a tall-cane Begonia which lives inside in winter and just outside in summer. Flowers are pink outdoors and fade to white inside.

Begonia **'Benitochiba'**: on the north-facing kitchen windowsill. Marmite plant, but I love it.

Hyacinth: 'White Pearl', 'Anastasia', 'Roman Blue' and 'Roman White', forced in stages and planted in the garden when they are over.

Narcissus: 'Paper White' or 'Zeva'.

Pelargonium **'Ardens'**: tuberous pelargonium with tiny scarlet flowers in spring and summer before a period of dormancy.

Pelargonium **'Betty Catchpole'**: a soft orange.

Pelargonium **'The Boar'**: I love a strong orange or red flower in winter.

Pelargonium capitatum: pink with rose-scented leaves.

Pelargonium **'Lady Plymouth'**: variegated silvery leaves and smells of lemon.

Pelargonium **'Mabel Gray'**: best lemon-scented leaves of all.

Pelargonium papilionaceum has huge heart-shaped lemon-scented leaves and unshowy purple flowers. The south-facing windowsill plants are an alternative to curtains.

Schlumbergera × *buckleyi*: Christmas cactus needs rescuing and resting after flowering. Likes shade in summer.

Sparrmannia africana: indoor lime, with huge furry leaves and white flowers. Very easy from cuttings and the indoor ones are allowed to grow to six feet outdoors in summer before being cut down to size and propagated again.

Allotment

Kale: Red Russian, 'Cottagers' and 'Cavolo Nero', all dependable in the winter months if they are protected from pigeons and deer.

Kalettes: nicer than Brussels sprout, I think.

Salads: less good since the deer found them. I'm going to grow many more next year. Chicory 'Variegata di Castelfranco' is a winter favourite, so is 'Rosso di Treviso'. Mustard 'Osaka Purple' also survives.

Squash 'Crown Prince': keeps all winter; if I could only grow one squash it would be this. 'Kuri' and 'Turk's Turban' look cheerful on the windowsill, but rarely keep for long and are less delicious than 'Crown Prince'.

Tree cabbage: a perennial cabbage which you treat like a kale, picking the large leaves when you need them.

Winter spinach 'Monstrueux de Viroflay': I saw in Dan and Huw's garden, and has been a great success. Some under a cloche, some not.

Spring Plants: March, April, May

Springs tend to be hotter now than they used to be, so some flowering times may vary, and some plants may continue into early summer.

Trees, Shrubs and Climbers

Akebia quinata: in the cream form which I now regret choosing instead of purple. The flowers are too tiny to register, but they are scented and this is a rampant climber for difficult places. Plants to cover the four arches over the paved passage from the courtyard to the rest of the garden have to share a bed with Bob's lonicera hedge. Two arches are *Akebia*-covered and two with the rose 'Adélaïde d'Orléans', which does surprisingly well.

Ceanothus arboreus **'Trewithen Blue'**: this tree form of the Californian lilac contributes to the informal evergreen screen between Bob and Lyn's house and my garden. Along with the rose 'Cerise Bouquet' and the silver-leafed Moroccan broom, it was planted to replace a row of leylandii.

Clematis alpina **'Willy'**: early flowering and has pretty seed heads, but it needs tying in to stop it becoming a tangle.

Crataegus coccinea: the scarlet hawthorn. I find it hard to resist planting hawthorns. They have two seasons of interest and the birds love them too.

Crataegus orientalis has silver leaves followed by coral berries.

Crataegus persimilis **'Prunifolia'** has larger leaves than other hawthorns.

Daphne laureola: the scented green-flowered woodlander.

Daphne odora: daphnes officially dislike being moved once established but I have had success moving this and *D.* 'Eternal Fragrance'.

Daphne × *transatlantica* **'Eternal Fragrance'**.

Jasminum polyanthum: the tender scented jasmine ought not to survive Cotswold winters, but it grows and flowers on the south wall of the house in the street. A hard winter sees it browned off but not yet killed.

Magnolia liliiflora: also inherited, this was hidden behind another leylandii barrier which we removed, but has grown well since then. 'Nigra', the dark form of this pink magnolia, is even more desirable.

Magnolia × *soulangeana*: we were lucky to inherit this, but the path out of the greenhouse had to curve to accommodate the tree. Late summer pruning is the only moment to keep a magnolia in shape.

Malus domestica **'Ashmead's Kernel'**.

Malus domestica **'Discovery'**.

Malus domestica **'Egremont Russet'**.

Malus hupehensis: I have two forms of this. The Dixter form has tiny berries. It will get huge so I'm trying it as a multi-stem which will need coppicing from time to time. The other tree from a reputable nursery has much larger flowers and almost apple-sized fruit.

Paeonia **'Bartzella'**: the intersectional Itoh peonies are perhaps better for small gardens than the *lactiflora* herbaceous kinds, because their leaves survive well after flowering. These tree peony crosses are showy and every year I wonder whether to remove 'Bartzella', which seems too large, too bright in the May garden.

Paeonia delavayi: a dark crimson form. This tree peony was in our last garden and seeded itself freely, in all shades from rust to almost black. One seedling moved with us. It is a plant I would always want to grow.

Paeonia rockii: my favourite tree peony. Its huge white dark-centred flowers are in scale with the size of the shrub and its leaves remain beautiful. I have two seed-raised *rockii*, both pale pink and equally lovely. It is a seven-year wait for home-grown seedlings to flower.

Paeonia **'Simply Red'**: another itoh cultivar, this has large flowers in a more forgiving colour than 'Bartzella'. It is a larger version of the species *P. delavayi* in the apple-tree bed across the path.

Philadelphus **'Belle Étoile'**.

Philadelphus coronarius: all the mock oranges are worth growing for a succession of scented flowers.

Philadelphus coronarius **'Aureus'**: I like this for its gold leaves in spring. The colour fades in summer. If it didn't, I would like it less. Pruning philadelphus after flowering is vital. My aim is long curving wands rather than twiggy growth.

Philadelphus **'Manteau d'Hermine'**: much smaller than the rest.

Philadelphus **'Mexican Jewel'**: smells of bubblegum.

Philadelphus **'Starbright'**: a new, very upright form with purple stems and shoots.

Prunus domestica: inherited damson.

Prunus insititia **'Mirabelle de Nancy'**: plum from Pippa's Somerset colony. It makes a tall, narrow tree and is self-fertile, but is late to blossom in this garden.

Prunus mume **'Omoi-no-mama'**: I grow this in a pot, but it can make quite a large shrub or tree.

Prunus persica **'Peregrine'**: in flower early so frosts can blight fruit buds, but its south wall position is now overshadowed by the reach of non-Dixter *Malus hupehensis*, so that may need rethinking.

Pyrus domestica: an inherited pear which has small inedible fruit, so I picked some blossom to bring indoors, only to discover that the smell is disgusting.

Pyrus salicifolia **'Pendula'**: the silver-leafed weeping pear tree was here when we came. The counsel of perfection suggests pruning it to leave separate trailing strands, as John Massey at Ashwood does. The tree is high, so this is difficult, but occasional thinning to allow enough light for plants below is important.

Rhododendron **'Fragrantissimum'**: reputedly tender, but survives here in a pot in the north-facing corner of the courtyard nearest the house.

Ribes sanguineum **'King Edward VII'**: the pink-flowering currant which we inherited here, but I would rather have *R.* 'White Icicle'.

Ribes speciosum: a plant which puzzles people. Its scarlet flowers look a bit like those of a fuchsia, so its common name is fuchsia-flowered gooseberry. It can be wall-trained. I use the prickly prunings on the flower beds to deter cats.

Rosa banksiae **'Lady Banks'**: the double white scented banksia which smells of violets.

Rosa xanthina: very early, with pale yellow flowers and ferny leaves. I grow it in memory of my father-in-law Hugh. *R.*

cantabrigiensis is similar but more upright where space is tight.

***Salvia rosmarinus* 'Sissinghurst Blue'**: this rosemary was a gift from Sue Dickinson, the Paradise and Plenty head gardener, who also worked at Sissinghurst.

Schisandra rubriflora: the Chinese magnolia vine, a strange twining woody climber, happy on a shaded wall, with tiny crimson flowers followed by berries.

Staphylea colchica has pendulous white flowers and bladdernut seeds. It is too big, so I have to coppice it, but as I grew it from Colesbourne seed, I want to keep it. The pink form is even prettier.

***Wisteria sinensis* 'Prolific'**: finally looks like flowering after four years of waiting.

Perennials and Bulbs

***Ajuga reptans* 'Catlin's Giant'**: the largest bugloss with dark leaves and blue flowers will cover shady ground fast. I use it in the passage out of the garden leading to Back Lane.

***Anemone blanda* 'Blue Shades'**: recommended for flower beds and rockeries by the RHS, but I have enjoyed it naturalised in grass, both in the last garden and here. It is happy in shade or sun. I never mind the occasional arrival of a white flower.

Anemone hortensis: scarlet anemones are among my favourite flowers and this hard-to-please Greek windflower comes and goes in the little meadow. I have tried pushing seed into the crack between the pavement and the south-facing house, which is as Mediterranean as you could get, but it will be years before I can see if that has worked.

Anemone × lipsiensis: a shade-loving pale yellow anemone which should spread if it settles in the cracks in the irregular pieces of stone in the small sunken garden.

Aquilegia chrysantha: another long-season aquilegia. I have grown different coloured forms of columbine in other gardens, but they tend to intermarry and end up as small doubles in murky shades. I once had a disastrous affair with *A*. 'Magpie' and ended up hating it and its progeny. There is a new fatal downy mildew disease which is affecting columbines. Plants showing signs of this should be dug out and destroyed.

Aquilegia longissima: this long-spurred lemon-yellow granny's bonnet flowers for ages if deadheaded.

Brunnera macrophylla: like a giant forget-me-not, is useful ground cover in shade.

***Brunnera macrophylla* 'Alexander's Great'**: a larger version of 'Jack Frost'.

***Brunnera macrophylla* 'Jack Frost'**: the variegated leaves are an eye-catcher.

Camassia quamash: the quamash is the smallest and bluest of the camassias. Officially it likes a damp meadow but it grows well here.

Convallaria majalis: lily of the valley is picky, but once happy it is unstoppable. I grow it in a dark corner of the courtyard, behind pots, under a high wall.

Cyclamen repandum: this lovely spring cyclamen is scented and likes damp woodland best. It does well in the West Country. I try to encourage it here under the *Prunus subhirtella*, not very successfully.

***Epimedium* 'Amber Queen'**: a recent purchase, with intriguing leaves and flowers.

Epimedium × *versicolor* **'Sulphureum'**: I have always been a bit resistant to *Epimediums*. Such a lot of leaf for so little flower and you have to remember to cut all the leaves off for a sight of the flowers. But this one is a useful ground cover for dry shade.

Erysimum cheirii **'Vulcan'**: a dark red wallflower which I often forget to sow. Bought plants tend to be slower than home-grown ones. Some wallflowers are perennial for a couple of years if you leave them in the ground, but mine tend to be planted in pots, which makes them harder to keep.

Erysimum **'Constant Cheer'** or **'Winter Orchid'**: I prefer these to the popular *E*. 'Bowles Mauve', because I find reds and oranges more uplifting than mauve, but it has to be admitted that 'Bowles Mauve' is the sturdiest of the pack. All the 'perennial' wallflowers need renewing from cuttings when they start to straggle.

Euphorbias are plants which I find impossible to ignore. The large spurges provide a strong presence, the small ones have unusual leaves, and all of them have lime-green flowers over many months. Their sap is lethal, so wear gloves when any part of the plant is cut and never rub your eyes after touching a euphorbia.

Euphorbia amygdaloides **var.** *robbiae*: useful for dark corners.

Euphorbia ceratocarpa: this tall, airy spurge from Sicily is my favourite. It likes to be hot and dry. A wet winter will kill it, but it strikes easily from cuttings.

Euphorbia characias **subsp.** *wulfenii*: a good accent plant. I have had named forms of this, such as 'John Tomlinson' and 'Lambrook Gold', but they are short-lived. Mine seed freely, so

I always have plants, but I occasionally get a black-eyed form which I like less than the pure green. Once the flowering stems start to fade they should be removed at the base to make way for new leaves.

Euphorbia donii **'Amjillasa'**: Fergus Garrett told me this was his favourite Euphorbia a few years ago. It flowers from late spring until autumn, but I may be growing it in over-rich conditions as it flops. Grown hard, it might be more upright.

Euphorbia griffithii **'Dixter'**: an exception to the lime-green norm. It has reddish stems, which run once it gets going, and bright orange flowers.

Euphorbia myrsinites: a low blue-green form which needs a dry hot place at the front of a bed.

Euphorbia seguieriana **subsp.** *niciciana* has needle-thin leaves and makes a low-spreading mound.

Fritillaria imperialis: I sometimes grow crown imperials in pots as they do not persist unless you leave them to die right down and in a small garden their unsightliness is hard to hide.

Fritillaria meleagris: the snake's head fritillary flourishes in damp meadows but also survives in drier places, like this garden.

Geum **'Totally Tangerine'**: long-flowering orange flowers for sun, but not always as persistent here as others claim.

Gladiolus tristis: this slender early flowering gladiolus is a far cry from the Dame Edna kind. I have grown a few in a pot and am now trying them in the courtyard. A clump of these evening-scented flowers in early April is hard to beat. But they do need support.

Irises do well here. I prefer the less complicated sorts, not too many colours and no frills. They get divided about every three years after flowering. I don't cut their leaves back after flowering, until they turn brown. This is unorthodox but the foliage is a great addition while it lasts and the practice of cutting them into fans looks a bit severe in this relaxed garden. I keep meaning to source one of the remontant irises to see if they really do flower a second time. 'Lovely Again' looks worth a try.

Iris **'Benton Nigel'**: a dark purple, which looks good with the dark stemmed cow parsley.

Iris **'Benton Old Madrid'**: a pinky brown gentle presence under *Rosa* × *odorata* 'Mutabilis'.

Iris **'Benton Susan'**: a weird brown with white falls, a present from a good gardening friend.

Iris flavescens came from Miserden Garden and has pale lemon-yellow flowers. If I could only grow one iris, it might have to be this very old and unimproved form.

Iris pallida: the Dalmatian iris, has leaves which look good all summer and pale blue flowers. I used to grow 'Jane Phillips' but taller large-flowered forms like this often need staking, so I tend to choose historic or species irises, which are shorter and less frilly than more modern types.

Iris **'Quaker Lady'**: this came from a garden called Daneway, where Vita Sackville-West helped the owner Oliver Hill to plant the garden. When I worked there I was given some of the smoky brown old-fashioned iris by later owners, the Spencers, when I was asked to restore their garden.

Iris sibirica **'Silver Edge'**: the Siberian irises are elegant with narrow leaves and easy to place almost anywhere. I like the blue-flowered forms best but they also exist in yellow and dark red.

Lathyrus vernus: a neat addition to the spring garden, for semi-shade I have dark purple and pink forms. It seeds in a different colour from the parent plant.

Lilium henryi is easier to please here than *L. martagon*. Emma Keswick has huge drifts of these Turk's cap lilies in her Cotswold garden, which she does admit are watered. The Madonna lilies I have not grown well. Regals come and go. Scarlet lily beetle is on the increase. You need to be quick to pinch them before they fall to the ground on their backs and become invisible.

Lunaria annua **'Corfu Blue'**: a more perennial form than the usual honesty and it is luminous blue in shade. Not purple.

Muscari armeniacum: in the garden when we came; *M. botryoides* is a bit classier. Grape hyacinths are despised for being easy and invasive, but they are early to flower and scented so they are allowed to run about at the back of the wall border. It is easy to pull them out if they become a nuisance.

Muscari armeniacum **'Valerie Finnis'**: porcelain blue and pretty in a pot.

Narcissi are the flowers of spring for most gardeners, but many of them are too bulky for this small-scale plot. I grow a few larger historic kinds in corners, where their dying leaves will not be too obvious. In the orchard meadow, where I want anemones, cowslips, primroses and tiny tulips to dominate, I am very selective. I often trial them in pots before planting them out.

Narcissus assoanus: a tiny jonquil. I am trying it on slightly raised sunny ground in front of the *M. liliiflora*.

Narcissus **'Bath's Flame'** has pale drooping flowers rather than brassy upright ones.

A LIST OF THE PLANTS I CURRENTLY GROW

*Narcissus **bulbocodium***: the hoop-petticoat daffodil is a favourite for pots.

*Narcissus **cordubensis***: an early species jonquil from Spain. It is said to like wet conditions, but after a very wet winter it failed to reappear.

*Narcissus **jonquilla***: the true highly scented jonquil, early to flower with thin leaves which die unobtrusively.

Narcissus **'Kokopelli'**: a scented jonquil type which I am trying to increase here. It has done much better in a garden in the south of France where I have recently been working.

Narcissus **'Lieke'**: I slightly regret planting this in the meadow as it looks clumsy among the small wildlings. Good in a pot and scented, so I will keep it to the edges of the garden.

Narcissus **'Minnow'**: a favourite for pots.

*Narcissus **moschatus***: a white scented species, with a drooping habit, from the Pyrenees. Too big for the meadow but beautiful.

*Narcissus **poeticus*** 'Recurvus': the pheasant's eye narcissus is taller and later than the rest in the meadow, but by the time it flowers the grass is longer, so it looks fine.

Narcissus **'Segovia'**: scented, multi-headed and the right scale for spring in the meadow here.

Narcissus **'White Lady'**: an old-fashioned pale scented daffodil for picking.

Ornithogalum pyrenaicum: you have to grow the Bath asparagus if you live near Bath.

Papaver cambricum: the Welsh poppy seeds yellow, or orange in shady places. There is a red form, 'Frances Perry', which is less persistent.

Primulas have made themselves thoroughly at home here, seeding in grass and flower beds. I am resistant to murky mauves, which occur when they cross with coloured forms. I used to grow the Barnhaven primroses in many shades of red and I miss them, but their promiscuity would be risky here.

Primula auricula: these tidy bright florists' flowers used to be my pride and joy, but I now grow very few. They are a lot of work, partly because root aphis are nearly impossible to banish and also because repotting them annually is a major undertaking. If you do want to grow them, the best advice is to keep them out of the rain in winter and out of the sun in summer. It is of course possible to grow some of the less fancy kinds in the ground.

Primula elatior: oxlips grown from seed are an elegant addition to the meadow.

Primula veris: the native cowslip, which has increased dramatically from Sue Dickinson's gift of an original nine clumps.

Primula vulgaris: I allow this to seed in the flower beds as well as in the meadow, but I remove plants in the flower beds to make way for half-hardies in May.

Primula vulgaris **'Hall Barn Blue'**: irresistible tucked in under roses but not allowed in the meadow.

Pulsatilla vulgaris: the pasque flower has beautiful seed heads. I prefer the dusky red form to the more usual mauve.

Ranunculus acris: the true meadow buttercup, much taller than the creeping *R. repens*, which was in the grass when we came, and *acris* carries on into summer.

Ranunculus acris **'Stevenii'**: a giant form of meadow buttercup which Christopher Lloyd gave me. I would like to grow the pale *R. acris* 'Citrina' again. I should sow some seed.

Rhinanthus minor: yellow rattle is the meadow maker's friend because it reduces the growth of grass, leaving room for flowers to grow. Rattle is an annual which needs to be sown fresh. Once you have it, you can harvest your own seed to sow in late summer. When the grass has been cut and carted, usually about the end of August but it could be later, scrape a few patches of earth until they are bare and then sow the rattle. It is better if the seed is in direct contact with the soil, but it can also sow itself into thin grass. In some years the rattle will dominate, but if the grass gets very sparse then the rattle population will decline. Unfortunately rattle will not subdue false oat grass, which has started appearing in the meadow. If this happens, clumps should be removed. Failing that, it must not be allowed to seed. Richard Barnwell recalls seeing people roguing the grass by hand in Cambridgeshire barley fields in the sixties.

***Saxifraga* 'Miss Chambers'**: a larger form of London pride growing in the darkest corner of the paved area.

Scilla bithynica: the Turkish squill, very early, very blue, increases well in shade. I first saw it at Great Dixter. It seems happy here.

Scilla luciliae: blue and white. I have introduced it in the flower beds as it flowers so early.

Scilla siberica: the most piercing blue of the scillas, easy anywhere but does not like to be very dry.

Silene dioica: the pink campion is usually out in May.

Silene vulgaris: the white bladder campion has edible leaves.

Smyrnium perfoliatum: has bright lime-green flowers and once established you have more than you want for ever, but it takes three years from seed to flowering.

Tulips: I have given up ordering en masse for borders, but I still grow some in pots and for picking. Those that I have enjoyed and might still use in pots are 'Sapporo', 'Prinses Irene', 'Rem's Favourite', 'Absalom', 'Carnaval de Nice' and 'Apricot Beauty', and I do regularly try others, but rarely choose large reds in spring. I like lily-flowered forms best and those tulips which I could never forgo are:

Tulipa **'Antoinette'**: a multi-stemmed small tulip in sunset shades. It starts yellow and fades to pink. Not too large, it fits in well here.

Tulipa **'Ballerina'**: scented, perennial, orange and elegant. Some more scented tulips are 'Generaal de Wet', which is now hard to find, and the yellow 'Bellona' and 'Striped Bellona'. I used to grow all these.

Tulipa clusiana: it is hard to source true *T. clusiana* now, but 'Peppermint Stick' is probably as near as you can get and it seems to be perennial in grass. I dislike the yellow hybrids on offer.

Tulipa 'Cornuta': I am trying this strange pointy-petalled species in the meadow and it does reappear.

Tulipa hageri **'Little Beauty'**: the tiny crimson flowers are easy to naturalise.

Tulipa linifolia **'Honky Tonk'**: the name is terrible, but the flowers are slender, lemon yellow with a hint of pink, and happy in grass.

Tulipa **'Queen of the Night'**: has lasted for years in a row for picking at the allotment so I will try some in the borders, behind other plants so that their dying foliage is hidden.

Tulipa **'Spring Green'**: lasts in the courtyard beds. Viridifloras do seem to be reliably perennial.

Tulipa sylvestris: a scented British native which spreads by underground stolons if it is happy, but others have reported virused strains. If leaves are distorted then bulbs need lifting and destroying.

***Tulipa* 'White Valley'**: I have enjoyed this in pots in the courtyard.

Zizia aurea: a gold version of the native alexanders, with bright yellow flowers.

Allotment

In the allotment there is:
Broccoli: Purple and white.
Kale: Curly and Red Russian.
Leek 'St Victoire'.
Lettuce: first picking mid-May.
Spinach.

Allotment flowers are:
Sweet william: Elite strain.
Tulips: 'Orange Favourite' and 'Queen of the Night'.

Summer Plants: June, July, August

Here there might be some overlap with spring. The time of flowering varies from year to year, with a few plants appearing in late May and others continuing into early June.

Trees, Shrubs and Climbers

Abutilon **'Ashford Red'**.

Abutilon **'Canary Bird'**: I grow the abutilons in pots and keep them in the shaded passage leading out of the courtyard. I take cuttings in case they fail to survive a hard winter and prune them hard at the end of spring.

Argyrocytisus battandieri: the Moroccan broom has bright yellow flowers which smell of pineapple. Its silvery leaves are the favourite food of snails, which is annoying. If it gets too big, I prune it after flowering.

Bupleurum fruticosum: the Mediterranean shrub is tiny in the wild, but unstoppable on better soil. Its green flowers attract masses of pollinators. I prune it quite hard in spring, otherwise it gets too big. At Great Dixter Christopher Lloyd regularly pruned it to the ground.

Cistus × *cyprius*: the gum cistus is not ideal here, because it leans out from under the fig to reach the sun, and the flowers now strike me as too white, but I continue to grow it for sentimental summer holiday reasons.

Clematis **'Alba Luxurians'**: works well here. I let it climb up the flowering currant on the boundary and into the mirabelle tree. I miss growing *C. Rehderiana*, which smells of cowslips, and the lemon peel *Clematis tangutica* with its silky seed heads.

Clematis texensis **'Gravetye Beauty'**: elegant, pendulous and crimson. Late flowering, but worth the wait. I used to grow larger blue hybrids like 'Perle d'Azur' and 'Prince Charles', but they are too showy for the wilder, airier look I aim for now.

A LIST OF THE PLANTS I CURRENTLY GROW

Fuchsia **'Enfant Prodigue'**: a showy hardy fuchsia which Christopher Lloyd liked, so that has earned it a place in the second bed outside the kitchen window.

Fuchsia hatschbachii makes a slender shrub with fragile flowers. It grows in the darkest corner of the courtyard in the bed under the kitchen window, so it never gets as large as it might in more favoured places.

Fuchsia **'Lady Boothby'**: a great addition to the courtyard wall, with non-stop flowers from summer until the frosts. I cut it down at the end of April. Hardy fuchsias are invaluable plants if you enjoy colour over a long season. I like, but do not grow here, 'Checkerboard', 'Hawkshead', 'Lady Bacon', 'Whiteknights', 'Whiteknights Pearl', 'Riccartonii' and *F. magellanica*, which is evergreen in mild climates. I enjoy a *F. magellanica* seedling from my neighbour's garden, which has landed on my side of the path to Back Lane. All of these have single, delicate flowers and will survive in part shade.

Gladiolus murielae: rarely perennial here, so I usually have to buy bulbs to enjoy the scent of these Abyssinian gladioli. It is important to choose a reliable source which will supply bulbs large enough to flower.

Lonicera etrusca **'Superba'** grows over the arbour and rose 'Francis E. Lester', but the better form of this honeysuckle is 'Michael Rosse'.

Lonicera implexa: Pip grew this evergreen Mediterranean honeysuckle from a cutting taken in France. Its leaves are blue-green and the flowers are scented. I must find a place for more honeysuckles. *L. periclymenum* 'Graham Thomas' flowers for the longest, but its flowers are pale creamy yellow and I prefer the pink-budded *L. periclymenum* 'Serotina' (the Late Dutch form), which flowers for longer than the earlier *L. periclymenum* 'Belgica'.

Roses need more water and nutrition than sustainable gardening allows, but species and single roses are less demanding than modern types and better for pollinators. Black spot is worse in wet seasons, but this garden is small enough to pick off affected leaves. Foliar feeding with seaweed fertiliser seems to deter aphids and in gardens rich in biodiversity, the greenfly will be devoured by birds and insects.

Rosa **'Adélaïde d'Orléans'**: recommended by Michael Marriott, the rose expert, for the arches over the passage. The roots of the roses are threatened by the lonicera hedge, as well as the occasional strand of ivy on the boundary with my neighbour, but 'Adelaide' still performs magnificently.

Rosa **'Albéric Barbier'**: my mother-in-law grew this rose, which is almost evergreen, with small creamy flowers, but it is unruly and grows fast on the back wall of the potting shed.

Rosa **'Albertine'**: a sentimental choice for its pink buds and heavenly scent.

Rosa **'Cécile Brünner'**: an elegant miniature tea rose which flowers all summer. It can be grown as a climber, but mine is a shrub. I like its airy growth and pointed pale pink flowers.

Rosa **'Céleste'**: an old-fashioned Alba rose with grey-green leaves and flat quartered soft pink flowers.

Rosa **'Francis E. Lester'**: has flowers like apple blossom and hips to follow. It grows over the arbour at the top of the garden.

Rosa **'François Juranville'**: scrambles into the overhanging branches of my neighbour's lilac. Its leaves are dark and glossy.

Rosa **'Geranium'**: the point of this for me is its height and its lacquer-red bottle-shaped hips. But the small blood-red flowers are pretty too.

Rosa **'Gertrude Jekyll'**: rather out of character with the rest of the garden. Showy rich pink flowers keep appearing, but I am not sure I will keep it. It is one of David Austin's best-loved roses, but it would be better in a more formal setting than this one.

Rosa **'Ghislaine de Féligonde'**: long-flowering apricot, can be grown as a climber or a shrub.

Rosa **'Guinée'**: deepest crimson scented rose with stiff stems. This is a challenge to grow. It needs plenty of compost and careful tying in so that the shoots do not snap. 'Étoile de Hollande' is much easier to please, but has larger paler roses than the mysterious 'Guinée'.

Rosa **'Madge'**: grown at Dixter and a favourite of Christopher Lloyd, its small scale makes it ideal for the little house at the end of the garden.

Rosa **'Munstead Wood'**: a modern shrub rose, now discontinued. I want to keep it going for the sake of its quartered deep crimson velvet flowers.

Rosa × *odorata* **'Bengal Crimson'**: a single cherry-red chinensis rose, so slightly tender. It was very slow to start, both here and in the last garden, but it seems now to have settled on the west-facing wall of the courtyard.

Rosa × *odorata* **'Mutabilis'**: one of my desert island plants. It flowers from May, throughout the summer and often produces the odd rose up to Christmas.

Rosa **'Paul's Himalayan Musk'**: if you want one rampant tree climber this is the one to have. 'Rambling Rector' is lovely too and both have the advantage of hips.

***Rosa* 'Phyllis Bide'** has clusters of repeat-flowering small roses. I have a weakness for plants which change colour as they age. The tulip 'Antoinette', the mutabilis rose and 'Phyllis Bide' all go through stages of salmon pink and yellow. Odd, because I used to despise the roses 'Peace' and 'Masquerade', perhaps because they were too large and too vivid.

***Rosa* 'Scharlachglut'**: a sensational plant with large crimson single flowers and showy hips to follow.

Perennials, Annuals and Bulbs

Acanthus mollis **'Rue Ledan'**: the white acanthus, undeniably handsome, but undeniably invasive and I slightly regret planting it, as it is very hard to remove once established.

Agapanthus **'Loch Hope'**: the hardy form I like best. It is an AGM plant. There are newer navy-blue varieties, if you like dark colours, and an intriguing recent long-flowering kind, the Everpanthus series, but these plants are shorter. Evergreen *Agapanthus africanus* is better for pots because it is doubtfully hardy. Mine spend the winter in the little house at the end of the garden and come out at the beginning of May. Agapanthus need plenty of water after flowering. Dividing pot-bound plants is a job for two people and a small saw.

Alcea rosea: the cottage garden favourite in single shades, not double. I have written on how to manage these at length in the July chapter.

Alcea rugosa: the lemon-yellow fig-leafed hollyhock, perennial for a few years if it is cut to the ground after flowering, and slightly less prone to rust than other hollyhocks.

Allium sphaerocephalon: the dark crimson drumstick allium. I used to major in alliums at the last garden, in particular *A. siculum*, with its strange drooping brown flowers, as well as the widely overplanted *A. hollandicum* 'Purple Sensation', but I resisted planting them again when we moved. I still like *A. cernuum*, so I grow that. There is a named form, 'Hidcote', which I would like to try.

Allium **'Summer Beauty'**: a late variety, which seeds itself in the apple-tree flower bed.

Aloysia citrodora: lemon verbena survives in the courtyard here. I have to prune it drastically after all danger of frosts has passed, or it shades the chinensis rose. The dried leaves are good for tisanes.

Althaea cannabina: one of my favourite plants, very tall and airy, like a tiny pink hollyhock, it flowers for months. Chelsea chopping it keeps it under control. Just.

Anchusa azurea **'Loddon Royalist'** has piercing blue flowers. Not perennial for me, but very easy from root cuttings.

Anemone hupehensis: an inherited pink form here is a thug and I would prefer to banish it. The white *Anemone* × *hybrida* 'Honorine Jobert' is useful for late summer and slightly more manageable, but I have not planted it in this garden.

Anemone hupehensis: a white species from Szechuan collected by James Compton in 1995. Like all wildlings, it is more delicate than any cultivar.

Anthriscus sylvestris **'Dial Park'**: a cow parsley with even darker leaves than 'Ravenswing'. It seeds about, but not as much as the native form. I allow one plant of the wild plant at the far end of the garden and behead it before it sets seed. At Dixter, cow parsley is threaded through the early summer borders. Beautiful, but very high maintenance is needed to stop it taking over.

Aquilegia longissima: my favourite columbine, will keep flowering into summer if deadheaded, but granny's bonnets dislike dry summers so they rarely stay long here. In damp West Country gardens I have seen them naturalised in grass and the double blues and pinks do sometimes appear in the meadow at the shady end. Downy mildew brought on by wet conditions shows as blotched and distorted leaves. There is no cure, so plants with signs of this should be destroyed.

Astrantia **'Roma'**: a fail-safe long-flowering plant in good rich moist conditions, which it does not often get in this garden.

Athamanta turbith: mimics a tiny cow parsley with bright green feathery leaves. It needs a hot, dry place.

Baptisia australis looks like a wild lupin with steely blue foliage and once established it is happy in sunny dry places. There are many cultivars of this plant in blue, yellow and pink, but I like the straight blue best. It is really better on acid soil, but it seems to cope here.

Calamintha nepeta **'Blue Cloud'**: like a tiny airy catmint. It flowers for months.

Campanula lactiflora **'Prichard's Variety'**: an old favourite which keeps going if deadheaded. I have tried Chelsea chopping alternate stems to prolong the season but am not convinced this is helpful.

Campanula portenschlagiana: evergreen and spreads rapidly in the dark, not very deep beds outside the north-facing kitchen window. I rip it up annually and put it back in its corner, but it is a useful plant for a difficult place and the flowers are pretty.

Campanula pyramidalis: the tall biennial chimney campanula, easily grown from seed. It will last for weeks in a cool room indoors. Outside, once pollinated, the flowers turn brown. It is possible to deadhead the plant, but this is time-consuming work.

Cenolophium denudatum: a fail-safe, self-seeding, more glamorous version of cow parsley. I like plenty of umbellifers and this is one of the easiest to please, happy in sun or shade.

Centranthus ruber **'Atrococcineus'**: the red form of the naturalised flower known as valerian. It flowers for months and if deadheaded, it will flower all summer. I dislike the murky pink form of this, but find the red form very pleasing in dry places like the little lane that leads from the door at the back of the garden. The white form is useful too.

Chaerophyllum hirsutum **'Roseum'**: pink chervil for a shady corner here. It can take more sun in damper gardens than this one.

Chamaenerion angustifolium **'Album'**: the white version of the common rosebay willow herb. I let it run along the hedge boundary under trees, but I would not put it in a border. The pink cultivars 'Stahl Rose' and 'Isobel' are even prettier. I have 'Isobel' down by the compost heap. The fluffy seeds are wind-blown, so it is probably best to remove them if you are alarmed by the idea of a willowherb takeover.

Cicerbita plumieri: a hefty sow thistle with blue daisy flowers, a gift from my Irish friend Christopher Moore. It's coarser than most of the plants I grow but beautiful. Needs support.

Cichorium intybus: blue flowers in the meadow from late summer into autumn.

Crocosmia × *crocosmiiflora* or **montbretia**: so invasive that it is now illegal to plant it in the wild. I still love it for its early gold green leaves and orange flowers, followed by seed heads.

Crocosmia **'Hellfire'**: I think this is a better plant than the larger 'Lucifer', which I used to grow, because its flowers are a deeper red and it does not need staking. *Crocosmia* 'Emily McKenzie' has big burnt-orange flowers but likes wetter conditions than I can provide. I admired a yellow *crocosmia* in Ireland, *C. masoniorum* 'Rowallane Yellow', which I tried in the allotment, but failed to please it.

Dactylorhiza fuchsii: after Aaron Bertelsen gave me one common spotted orchid, it was a thrill to find another one, followed by four more appearing in the meadow. The seed of the spontaneous arrivals might have been dormant, or wind-blown. In time, the population should increase without any help from the gardener. I am not sure whether the bee orchids, another present from a kind friend, have survived, but they and the green winged orchid are established in a field at the top of the village, so I am hoping they may be blown this way on the prevailing west wind.

Dahlias are a summer mainstay. I like single forms best, but the Karma series are the ones to choose for picking, because they have been bred to have long stems. A few dahlias, 'Waltzing Mathilda', 'Magenta Star', 'Honka Fragile' and 'Verrone's Obsidian', appear every year. *Dahlia coccinea* 'Mary Keen', a tall, graceful plant with single crimson flowers, was a lucky find at a time when Avon Bulbs were trying to breed a black-flowered form. Those that were rejected were sold off at cut price. I liked the one they sent me and showed it to Christopher Lloyd, who also thought it was a good plant and he got it named. It survives the winter with a covering of compost under cut branches of evergreenery. Yew or thuya both offer good protection. Another species dahlia is *Dahlia merckii*. This has tiny flowers of white or mauve and looks good in a pot. I am attached to two shocking-pink dahlias, 'Winston Churchill' and 'Karma Fuchsiana', as well as to the pale pink 'Porcelain', while a new discovery is the glamorous Hadrian series from Halls of Heddon. Each autumn some tubers are dug up and put into pots with dry soil, after they have first been upended so that they dry out completely. They stay in the greenhouse and get watered in March, but after that they are better kept dry until the shoots appear. Dahlias in the ground are likely to be attacked by slugs in wet springs. The Bishop series is easy from seed and will flower in a season if you start early enough.

Daucus carota: wild carrot is a feature of the late summer meadow. It climaxes in some years, when I cull some plants.

Delphinium elatum: grown from seed from Franklin Farm. I prefer the more delicate Belladonna series to the giant columnar delphs which are out of scale with this garden. 'Cliveden Beauty' is sky blue and repeat flowering if cut down after the first flush and watered. 'Völkerfrieden' is an intense dark blue.

Deschampsia cespitosa 'Goldtau': a small grass with shimmering flowers. I find grasses hard to integrate in more traditional flower beds, but this is pretty from late summer and through the winter.

Dianthus carthusianorum: the ever-flowering tall pink with tiny magenta flowers. One of my favourite plants, like all pinks, it needs replacing every three or four years, but cuttings strike easily. June is a good month to pull what are known as pipings, from shoots which are not flowering, and these should take fast in gritty soil round the edge of a pot. In the past I have had crazes for different dianthus like 'Gran's Favourite' or 'Laced Monarch' or 'Brympton Red' and the smart white 'Haytor', and I may do so again.

Dianthus cruentus: neater, smaller and brighter than *D. carthusianorum*.

***Dierama* 'Blackbird'**: the angel's fishing rod which I like best, but I also have a pink unnamed form. Christopher Lloyd grew them between paving stones. South African plants, they need a place in the sun, in well-drained soil.

Digitalis ferruginea makes tall spires of rusty brown. These are nearly as satisfying as the foxtail lilies which I consistently fail to please. Eremurus, the desert lily, likes sunny open spaces with plenty of room to breathe. It does not seem to want to rub shoulders with other plants.

***Digitalis purpurea* 'Sutton's Apricot'**: a biennial which needs sowing no later than June for flowers the following year.

***Echinops ritro* 'Veitch's Blue'**: a useful but bit underwhelming globe thistle. It's good for pollinators.

Erigeron annuus: a tall fleabane which makes cloudy-white splashes in the borders in late summer and autumn. It seeds everywhere, welcomed by me but not by more serious gardeners. There is a difficult moment when the forget-me-nots and some of the Corfu honesty must come out and the fleabane has to be redistributed, because it tends to seed on the edge of the beds into the path and large plants rarely survive moving. I like a matrix of one plant running through a flower bed to unite it. The fleabane performs the same unifying role as first primroses, then forget-me-nots, followed by honesty and then various umbellifers earlier in the year.

Erigeron karvinskianus: the South American fleabane is a delicate pink-tinged daisy which grows in the driest places and flowers from May until the end of summer, often into autumn. Here it grows on the south side of the house where the pavement meets the building. It's a good choice for the chinks in sunny steps or paving.

Eryngium giganteum **'Miss Willmott's Ghost'**: not perennial but seeds itself freely.

Eryngium × *olivierianum* **'Big Blue'**: a striking sea holly. 'Jos Eijking' is daintier. Hot and dry are the conditions to offer the sea hollies.

Euphorbia ceratocarpa: a tall, tender euphorbia in flower all summer. It needs careful pruning in late spring when a third of the older flowering wands should be cut back to new shoots. Spring cuttings work if you seal the shoots in warm water first.

Euphorbia seguieriana **subsp.** *niciciana*: another tall, willowy spurge with lime-green flowers all summer.

Ferula communis: the giant Mediterranean fennel grows next to the mimosa in the corner of the courtyard. It was a present from Dan Pearson when we moved and has only flowered once, but its leaves are beautiful, before they vanish in summer.

***Foeniculum vulgare* 'Purpureum'**: cutting fennel down in August will give it a second lease of life.

***Francoa sonchifolia* Rogerson's form**: an indestructible evergreen, often grown in a pot for its old-fashioned bridal wreath flowers of pink and white. It comes through the winter here outside, emerging with a few crisp leaves which are easy to pick off. I must try it in open ground where it would probably survive even better than it does in pots.

Galium album: the white flowers of hedge bedstraw are more in evidence than the yellow ones of lady's bedstraw, which is something I would like to alter.

Galium odoratum: sweet woodruff was once a strewing herb. Happy to spread in shade, it is ideal for the edge of the passage to Back Lane behind the garden.

Geraniums are excellent plants when you start a garden, or if you want trouble-free ground cover. I use a few at home, but try to keep them from spreading too much. In clients' gardens it is tempting to resort to G. 'Rozanne', because it flowers all summer in a blue that appeals to everyone. I used to grow G. *psilostemon*, which needed staking, but the smaller form 'Patricia' has equally large flowers and a longer flowering period, so that is a better choice. Sugar pink 'Mavis Simpson' has silvery leaves and flowers for months. Less well known, G. *wallichianum* 'Buxton's Variety' is a late summer blue with a white centre; I like the way its trailing stems climb into other plants. I can't have too many seedlings of the annual G. *pyrenaicum* 'Bill Wallis'. It seeds in the courtyard gravel and its tiny violet flowers are good in a pot with the pelargonium 'Copthorne'. When 'Bill Wallis' starts to get leggy, I cut the plant back to base and let it start all over again.

Geum 'Totally Tangerine': not as reliable here as others suggest. It flowers a little in early summer, but never for long, even if I cut it back.

Gladiolus communis* subsp. *byzantinus: it is increasingly hard to find the true form of this wild gladiolus, in what Christopher Lloyd called 'brilliant dashing magenta'. Many bulb nurseries offer a paler version which is sometimes *G. illyricus*, but if you want the zingy real thing, go to a guaranteed supplier like Great Dixter, or Beth Chatto. It will not seed, but increases by clusters of little bulbs around the parent.

***Gladiolus* 'Ruby'**: very desirable and it seems to be hardy. 'Ruby' bulks up well for Dan Pearson, but mine are not yet abundant. Sun in summer and a winter mulch will keep the corms happy.

Hemerocallis altissima: a tall, lemon-yellow day lily. The value of day lilies, for me, is that their fresh green leaves appear early. I used to grow brick-red 'Stafford', and pale yellows 'Marion Vaughan' and 'Whichford', but I found the daily deadheading – it's in the name – tiresome. When they regularly developed swollen buds due to gall midge, I gave them up.

Knautia arvensis: the meadow scabious is lovely in late summer but I do cull a few plants if too many seed, as it is large and can become overwhelming.

Knautia macedonica: the crimson pincushion scabious keeps on flowering all summer if it is deadheaded. There seem to be two forms around. I prefer the taller one.

Kniphofia thomsonii: a slender, elegant poker, pale orange and flowers for months, but it is slightly tender.

Laserpitium siler: king of the umbellifers with huge flower heads which turn pink with time. Beautiful at all stages of its life for a dry place.

***Lathyrus odoratus* 'Matucana'**: the best scented sweet pea is grown at the allotment, but I keep some plants in the garden to fill gaps in pots or flower beds, so that I can pick a few flowers at short notice.

Lathyrus sylvestris: a scrambling everlasting pea with modest pink flowers. I chop it to the ground in winter and then let it rise to climb the stems of the rose 'Guinée'. It is the fodder plant of the short-tailed blue butterfly.

Leucanthemum vulgare: the ox-eye daisy is dominant in the first phase of making a wild flower meadow. I wish I had the slightly daintier British native form rather than the thug which has taken up residence here.

Ligusticum lucidum: the shining lovage with white lacy flowers. I don't risk eating it, which is annoying because my still infant edible lovage is regularly devoured by slugs.

***Linaria* 'Canon Went'**: the pink perennial toadflax is good in the courtyard gravel.

***Linaria* 'Peachy'**: taller than 'Canon Went', in two-tone pink and yellow, and it flowers all summer in the courtyard.

Linum narbonense: a graceful bright blue flax. Difficult to grow from seed and not widely available.

Mathiasella bupleuroides: another green-flowered plant, bit Marmite, but I like it and it lasts for ages.

Matthiola incana: the perennial stock is a grey-leaved sub-shrub in scented flower for weeks, but not long-lived. Easy from seed or cuttings.

Molopospermum peloponnesiacum: a bright green umbellifer with ferny leaves happy in shade outside the kitchen window.

Nepeta nuda **'Romany Dusk'**: I saw this at Dan Pearson's garden, planted under *Rosa glauca*, and have copied it shamelessly here.

Oenothera stricta **'Sulphurea'**: the pale yellow scented evening primrose is a short-lived perennial, but good at reproducing itself in dry sunny places. It usually chooses the gravel path up to the greenhouse.

Oenothera lindheimeri: the gaura will flower all summer in a hot, dry spot. There are pink forms as well as white. Easy from seed. It has been a star performer in a garden we worked on in the south of France.

Origanum laevigatum **'Herrenhausen'**: very long-flowering marjoram and its seed heads last well into winter.

Oryzopsis miliacea: tall and airy grass which is lovely to pick in winter.

Paeonia **'Buckeye Belle'**: the only double peony here. I grow it in a dark group with 'Dial Park' cow parsley, under the rose 'Munstead Wood'.

Paeonia lactiflora **'Nymphe'**: single, scented, apple blossom pink. It prefers some shade but competition from the roots of the apple tree is not helping this plant, so I must find a better place for it to grow.

Paeonia mlokosewitschii: the true one is yellow and fleeting, but when I grew them from seed in the last garden they appeared, after three years, in all shades from pink through apricot, to yellow.

Paeonia obovata: a Japanese woodlander, single white and not very tall but a rare presence when it flowers.

Papaver dubium **subsp.** *lecoqii* **'Albiflorum'**: the lengthy name for the elegant pink Beth's poppy. I have seen this flowing through a Dixter border in early May and it's an effect I really want to copy. I have tried direct sowing and spring sowing into cells and then planting out, but poppies are tricky to separate and transplant. At Dixter they sow in autumn and overwinter the plants in a frame, then set them out in early spring in naturalistic drifts.

Papaver somniferum: does sow itself in a clear coral red and a dark purple. I remove any that are mauve. The keepers are singles; I am less enthusiastic about double forms.

Paris quadrifolia: grew in large colonies in the wood at our last place. Tucked into a hidden shady corner, with its green topknot of flowers, it seems to be quietly spreading.

Penstemon **'Sour Grapes'**: blueish purple and long-flowering. It is sometimes confused with P. 'Stapleford Gem', which is much paler. Penstemons are useful, but I only grow 'Sour Grapes' now. It is mostly hardy if stems are not cut back until late spring or early summer, but worth taking a few cuttings in case it fails.

Perovskia atriplicifolia **'Blue Spire'** (*Salvia yangii* **'Blue Spire'**): the Russian sage, needs cutting back before summer or it will become lanky.

Phlox paniculata **'Blue Paradise'**: phlox are beautiful in Ireland in moist, rich soil, so this plant needs a good mulch and sometimes water in free-draining gardens. I like the loose Dixter phlox known as 'Long Border Mauve' and wonder about trying Sarah Raven's new to me 'Fashionably Early Lavender Ice'. She has a good eye for new plants.

***Rotheca myricoides* 'Ugandense'**: not hardy but is lovely in a pot in summer in the courtyard, with blue flowers like butterflies. It copes with semi-shade. I will take a cutting when it gets too big, to replace it with a smaller plant. I wish I still had *Plumbago auriculata*, another blue-flowered beauty from Africa, which deserves greenhouse space in winter, but I failed to propagate its replacement. We used the miniature plumbago *Ceratostigma plumbaginoides* in the south of France where it flowers in the hottest, driest places. I used to grow the shrubby *Ceratostigma willmottianum*, but have not found a place for it here. All these are worth growing for their intense blue flowers in late summer.

Salvias have transformed summer gardens, so I grow plenty of different ones. Not all are hardy, but cuttings strike easily. I usually take these at the end of August. Some salvias will survive winter outside, but new shoots that grow from the base are slug fodder.

***Salvia* 'Amistad'**: hardy but not long-lived. It has become a bit of a garden cliché, because it is so easy to please and there is now a pink form of 'Amistad'. Like all salvias, it is better to leave the stems standing in winter and only cut them back when you see new growth. This salvia, according to Jonny Bruce, appreciates being watered.

Salvia confertiflora: dramatic and orange-flowered, but not hardy here. I find plants need to be overwintered as large specimens to perform. Cuttings never quite make it in time. *Salvia involucrata* 'Bethellii' is another large salvia with shocking-pink flowers, which will survive outside but needs a warm start to flower properly.

Salvia curviflora: tall with brightest pink flowers. I first saw it beautifully grown at Gravetye Manor. It would survive in a London or coastal garden, but will not overwinter for me.

Salvia darcyi: this soft green-leaved salvia with large scarlet flowers is tender and its stems are fragile, so give it the slightest knock and it will break. Best in a pot, although I did plant one in the courtyard which survived the last mild winter.

***Salvia microphylla* 'Cerro Potosi'**: a bushy, woody shrub that's as tough and hardy as you can get. Shocking-pink flowers all summer. Other useful *S. microphyllas* are the pretty 'Wine and Roses' and yellow forms. 'Hot Lips' I like less.

***Salvia* 'Nachtvlinder'**: a small hardy evergreen with dark delicate flowers all summer.

***Salvia patens* 'Guanajuato'**: tall with gentian blue flowers. Hardy for me, but not reliably so in cold gardens.

***Salvia* 'Phyllis Fancy'**: another tall Gravetye beauty with misty blue-grey flowers. Semi-hardy but lovely in pots.

Salvia sagittata: the arrow-leafed sage has stunning blue flowers but is not hardy.

***Salvia sclarea* var. *turkestaniana* 'Vatican Pink'**: sometimes perennial, but as a biennial it self-sows gently. Its crinkled grey-green leaves are handsome and its flowers showy, but their scent is disgusting. There is also a white form.

***Sanguisorba* 'Red Thunder'**: a giant burnet which keeps its red tassels for months, but it does fall apart here, perhaps because it prefers moister soil than it gets, so I have to stake it. Dan Pearson's invaluable Dig Delve blog recommends an iron girdle in mid-May. Once something has flopped, it is too late to start staking.

A LIST OF THE PLANTS I CURRENTLY GROW

Scabiosa columbaria **subsp.** *ochroleuca*: with little lemon-yellow flowers which Beth Chatto describes as 'dainty', it dances along the edge of the apple-tree bed for a good two months.

Silene coronaria **'Blood Red'**: darker than the usual magenta and just as good at self-sowing. I do deadhead it to keep it going all summer, only allowing it to seed at the end of August.

Succisa pratensis **'Derby Purple'**: dark form of the devil's bit scabious. It seeds around.

Thalictrum **'Elin'**: very tall but narrow, with lacy blue-mauve leaves. Like all the meadow rues, it does well in moist, fertile soil. I have two plants, one from Dixter and one from Special Plants. The plant in shade does less well than the one in sun.

Thalictrum **'Splendide White'**: a beautiful plant but it needs richer, moister conditions than I give it.

Verbascum chaixii: a perennial mullein, easy to keep unlike the silver-leaved giants *V. bombyciferum* or *V. olympicum*, which are beautiful but short-lived and caterpillar-prone. The lemon-yellow biennial *V. roripifolium* is my favourite mullein and it flowers for months, so it is worth the effort of annual seed-raising.

Verbena hastata **'Blue Spires'**: a charming branched verbena which also comes in pink and white shades.

Verbena officinalis **'Bampton'**: I like this short, airy verbena better than the overused *Verbena bonariensis*. It has dark leaves and self-seeds a little.

Allotment

In the allotment I spend less time than I ought now, so it's mainly for salads and more perennials or unusual sorts of veg and some flowers to cut. I would like to be more adventurous with modern methods so that work is easier, but making the transition from 'no dig' to cover crops is a big leap.

Asparagus 'Gijnlim'.
Artichoke 'Gros Vert de Laon': getting a named variety of the globe artichoke matters, otherwise plants grow into prickly, inedible heads.
Beetroot 'Chioggia': striped with white rings like a bullseye.
Broad Bean: 'The Sutton' or 'Crimson-Flowered'.
Carrot 'Flyaway': fly-resistant. I also like growing a few 'Purple Sun'.
Courgettes: only three plants or keeping up with the harvest is impossible. Usually yellow.
French Bean 'Blue Lake'.
Kale 'Red Russian': the small leaves are good in salads.
Lettuce: I rely on 'Morton's Secret Mix' for picking outer leaves. Sowing this three times a year is more reliable than sowing other varieties and picking the whole head. I do still sow 'Little Gem' as a first pick in May, and home-grown this lettuce tastes sweeter than any bought variety.
Peas: only mangetout.
Potatoes: only one row each of a first early and 'Charlotte'.
Spinach: inclined to bolt in our increasingly dry climate, but spinach beet survives and so does perpetual spinach, but I want

to try the pink *Chenopodium giganteum*, which seeds itself everywhere and is an ornamental addition to the plot.

And there are some flowers:

Cosmos **'Dazzler'** or **'Bright Lights'**: not too many, or picking becomes a burden.

Dianthus **'Electron'**: these sweet williams are perennial and auricula-eyed.

Lathyrus odoratus **'Matucana'**: this sweet pea has the best scent of all.

Nigella damascena: love-in-a-mist is a weed in the garden and the allotment. I plan to strim the band of them in the allotment this summer, before the flowers seed, and then try planting beetroot or courgettes among them in an attempt to adopt Josh Sparkes' techniques.

Zinnia elegans: I can't resist a few, especially 'Envy' and a mix of giant brights.

Autumn Plants: September, October, November

Autumn is the time when many of the summer plants carry on flowering and when spring-flowering trees bear berries and fruit, so this section will appear to have fewer named plants in it because many of them have been listed earlier. *Crataegus* and *Malus*, including the domestic kinds, all come into their own in September and October. Dahlias, salvias and fuchsias keep flowering until the worst frosts strike. Many roses have a second season, some with glamorous hips rather than flowers, and regularly deadheaded annuals also keep going long after summer ends.

Trees and Shrubs Specifically for Autumn

Abelia × *grandiflora* **'Francis Mason'** has glossy, semi-evergreen leaves and pinky white flowers which are slightly scented. I grow it in the darkest place in the garden, in the corner where the bins live. Dependable and beautiful, it deserves to be more widely grown.

Euonymus europaeus **'Red Cascade'**: a smarter version of the native spindle. I love its scarlet-and-orange fruit and the leaves colour well too. There are two *Euonymus* bushes on the boundary with my neighbour here, in front of the Portugal laurel. A straight-line boundary at the edge of any garden is something I try to avoid. Informal planting makes a town garden seem larger.

Ficus carica: here when we came, on the east-facing wall and very vigorous, so it gets a hard prune in spring. If you prune in summer you lose next year's crop of figs. 'Brown Turkey' is probably the most popular fig.

Heptacodium miconioides has very late white scented flowers and good autumn colour. It was here when we came, but much too tall, so I cut it back ruthlessly in early summer and it was unharmed. It would be better grown as a multi-stem tree rather than as a pollarded standard. *Heptacodium* would probably not have been my first choice for an ornamental tree in a garden of this size. Instead I might have chosen the hop tree, *Ptelea trifoliata*, which has cloudy green fragrant flowers followed by clusters of green-winged seeds.

Hydrangea paniculata **'Limelight'**: another inherited plant which I might not have chosen, but it grows against the wall in a shady corner, now almost smothered by *Clematis alpina* 'Pink Willy'. I cut the clematis back after flowering if I remember.

Perennials, Bulbs and Annuals

Aconitum carmichaelii **'Kelmscott'** has showy hooded deep blue flowers on tall stems. The monkshoods are happy in shade and enjoy damp soil, so 'Kelmscott' needs a heavy mulch here. The plants are poisonous.

× *Amarine tubergenii* **'Emanuelle'**: a cross between an amaryllis and a nerine, produces large pinky-white flowers and is completely hardy here.

Anemone hupehensis starts flowering in late summer, but autumn is peak season. The flowers are lovely but they are such an invasive tribe that I would never choose to grow them in a small garden where a wide variety of plants was the aim.

Chrysanthemum **'Dixter Orange'**: the only chrysanth I grow in the garden, but I grow several at the allotment and a few spider types in the greenhouse. I do lift the Dixter form and keep it under cover in winter, so that it can go out at a size which slugs will ignore.

Colchicum autumnale **'Album'**: the white autumn crocus, tricky to manage in very long grass. I halve the height of the grass where they grow before mid-August, so that the stunning white goblets can be seen. The large leaves appear in spring but disappear in summer, so you need to keep track of where the bulbs are planted.

Cyclamen hederifolium: the autumn-flowering cyclamen, which appears in Mediterranean countries as the first rains fall, is spreading in patches at the far end of the garden.

***Hesperantha coccinea* 'Major'**: bright red late-flowering flag lily from South Africa. I have seen it growing particularly well in gravel in a Bath garden. There is a pink form, 'Mrs Hegarty'.

Rudbeckia deamii has little orange daisies with dark centres. Best in sun and fine for dry soil. I like *R.* 'Prairie Glow' best of the coneflowers, but it is short-lived.

***Rudbeckia subtomentosa* 'Henry Eilers'**: a narrow-petalled pale yellow coneflower. It copes with dry soil and semi-shade under branches of the large apple tree.

Symphyotrichum (the terrible new name for 'aster') ***lateriflorum* 'Prince'**: low-growing and structural with dark leaves, although 'Lady in Black' is probably better as it has dark stems too.

***Symphyotrichum* 'Little Carlow'** used to be my favourite Michaelmas daisy. I still grow it, but it is stiffer than the graceful *S. turbinellum*, which I now like best of all for its starry blue flowers. 'Little Carlow' is first to flower.

***Symphyotrichum turbinellum* 'Leaflet'** has narrower leaves and slightly mauver flowers than *S. turbinellum*. Very pretty and airy, a present from Hannah Gardner.

***Symphyotrichum* 'Vasterival'** has cloudy pink flowers on dark stems. Taller than most, it can wait its turn behind other plants.

A LIST OF THE PLANTS I CURRENTLY GROW

Allotment

Here it's harvest time:

Beetroot: second crops.

Celeriac: if I have managed to water it enough.

Courgettes: so many you can't give them away.

French beans: Climbing 'Blue Lake' and some dwarf beans, 'Maxidor', yellow, or 'Aiguilon', a very thin variety. I no longer grow runner beans.

Kale: 'Redbor' and 'Red Russian' are still useful; 'Cavolo Nero' is starting to mature.

Lettuces: still in season, but sowings of winter salads and mizunas must replace them.

Raspberries: 'Autumn Bliss'.

Squashes: harvested now to keep on a sunny windowsill for winter. 'Crown Prince' is far and away the most delicious, while 'Kuri' is more ornamental. I like spaghetti marrow and early butternut.

Tomatoes: grown in the greenhouse, currently 'Gardeners' Ecstasy' and 'Primabella' from Real Seeds. This company regularly sources the tastiest veg.

And also for flowers:

Chrysanthemum: 'E. H. Wilson', 'Ruby Mound', 'Emperor of China', 'Allouise' sometimes, but 'Allouise' is not hardy.

Cosmos: might be 'Rubenza', 'Dazzler' or 'Bright Lights'. I like to vary what I grow from year to year.

Dahlias: 'Gerrie Hoek' lives at the allotment. It is too showy for the garden but useful for picking. I sometimes try other dahlias here, to see if they are garden-worthy.

Zinnias: 'Envy' and a mix of the Benary's Giant strain. They need a hot summer to perform well and it is a mistake to sow them too early. May is not too late.

Acknowledgements

My thanks are due to all those who helped me to write this book, in particular Natasha Fairweather, best of agents, and Nicholas Pearson at John Murray for believing in the diary from the start.

Thanks to my early readers and encouragers: Olivia Laing, Juliet Nicolson, Isabel Bannerman, Catherine FitzGerald, Christopher Moore, Catherine Goodman, Jane Smiley, Sheila Pearson.

And to the gardeners who continue to inspire me: Pip Morrison, Fergus Garrett, Tom and Sue Stuart-Smith, Derry Watkins, Sarah Raven, Sarah Price, Dan Pearson, Aaron Bertelsen, Jane Barnwell, Kirsty Knight Bruce, Joshua Sparkes and Peter Dennis.

Thanks to Ben Dark, whose words I quote at the beginning of this book. And for checking the botanical Latin, thanks to James Compton.

For invaluable help with picture research, thanks to Philippa Lewis. And from John Murray, Juliet Brightmore and Sara Marafini. Thanks also to Eva Nemeth for her photography, and to Joe Oswald for the garden plan.

For everything, always, thanks to Laura Beatty, Ellie Keen, Alice Oswald, William Keen.

Credits

Text

R. Kandeler and W. Ullrich, 'The Symbolism of Plants', *Journal of Experimental Botany* vol. 60 no. 2 (2009): 6–8.

Illustrations

January, August, September, October: wood engravings from *Four Hedges: A Gardener's Chronicle*, by Clare Leighton, 1935. © 2025 Estate of Clare Leighton. All Rights Reserved, DACS.

November, December, February, March, April, May, June, July and pages v, 184, 239: wood engravings and drawings by John Northcote Nash. © Estate of John Northcote Nash. All rights reserved 2025.

Garden plan, page x © Joe Oswald.

RAISING READERS
Books Build Bright Futures

Dear Reader,

We'd love your attention for one more page to tell you about the crisis in children's reading, and what we can all do.

Studies have shown that reading for fun is the **single biggest predictor of a child's future life chances** – more than family circumstance, parents' educational background or income. It improves academic results, mental health, wealth, communication skills, ambition and happiness.[1]

The number of children reading for fun is in rapid decline. Young people have a lot of competition for their time. In 2024, 1 in 10 children and young people in the UK aged 5 to 18 did not own a single book at home.[2]

Hachette works extensively with schools, libraries and literacy charities, but here are some ways we can all raise more readers:

- Reading to children for just 10 minutes a day makes a difference
- Don't give up if children aren't regular readers – there will be books for them!
- Visit bookshops and libraries to get recommendations
- Encourage them to listen to audiobooks
- Support school libraries
- Give books as gifts

There's a lot more information about how to encourage children to read on our website: **www.RaisingReaders.co.uk**

Thank you for reading.

---

[1] OECD, '21st-Century Readers: Developing Literacy Skills in a Digital World', 2021, https://www.oecd.org/en/publications/21st-century-readers_a83d84cb-en.html

[2] National Literacy Trust, 'Book Ownership in 2024', November 2024, https://literacytrust.org.uk/research-services/research-reports/book-ownership-in-2024